Beyond Belonging

Social Science for Social Justice Series

Social Science for Social Justice is a book series from Sage that provides a platform for academics, journalists, and activists of colour to respond to today's pressing social issues.

The series challenges the Ivory Tower of academia – in which black, Asian and minority ethnic voices are underrepresented – by defining the 'expert' not as someone who extracts data from a community but as someone who works within and alongside communities, gives back and amplifies voices. The series is interdisciplinary and international in scope and provides rigorous analysis and radical thinking which works towards a shared social justice which benefits all.

Books in this series include:

Consuming Crisis: Commodifying Care and COVID-19 by Francesca Sobande

The Muslim, State and Mind: Psychology in Times of Islamophobia by Tarek Younis

Whiteness, Racial Trauma, and the University by Harshad Keval

Queering the Asian Diaspora by Hongwei Bao

Slippery Eugenics: An Introduction to the Critical Studies of Race, Gender and Coloniality by R. Sánchez-Rivera

David Kam 甘家伟**, Amy Phung** 馮金薇
and Mai-Anh Vũ Peterson

Beyond Belonging

East and Southeast Asian Presence, Identity and Activism in the UK

3rd Floor
HYLO
103–105 Bunhill Row
London, EC1Y 8LZ
UK

2455 Teller Road
Thousand Oaks
California 91320

10th Floor, Emaar Capital Tower
2 MG Road Sikanderpur, Sector 26
Gurugram, Haryana - 122002
India

8 Marina View Suite 43-053
Asia Square Tower 1
Singapore 018960

Editor: Delayna Spencer
Commissioning Editor: Rhoda Toweh
Assistant editor: Harry Dixon
Production editor: Rabia Barkatulla
Marketing manager: Maria Omena
Cover design: Wendy Scott
Cover illustrations: Emma Yuan
Typeset by: C&M Digitals (P) Ltd, Chennai, India
Printed in the UK

© David Kam, Amy Phung and Mai-Anh Vũ
Peterson 2026

Apart from any fair dealing for the purposes of
research, private study, or criticism or review,
as permitted under the Copyright, Designs and
Patents Act, 1988, this publication may not be
reproduced, stored or transmitted in any form,
or by any means, without the prior permission
in writing of the publisher, or in the case of
reprographic reproduction, in accordance with
the terms of licences issued by the Copyright
Licensing Agency. Enquiries concerning
reproduction outside those terms should be
sent to the publisher.

**Library of Congress Control Number:
2025936207**

**British Library Cataloguing in Publication
data**

A catalogue record for this book is available
from the British Library

ISBN 978-1-5296-7320-3
ISBN 978-1-5296-7319-7 (pbk)

CONTENTS

For Peiwen
佩文

ABOUT THE AUTHORS

David Kam is a Malaysian movement artist, facilitator, speaker and founder of kindredpacket. He develops and shares embodied practices of freedom, care and connection by queering the frameworks of yoga, dance and architecture. Collaborating with leading dance and wellness institutions across the UK and beyond, David moves intergenerational communities towards reclaiming agency, celebrating expression and fostering collective wellbeing.

Amy Phung is a British Chinese writer, designer, speaker, community organiser and co-founder of besea.n, Britain's East and South East Asian Network. Drawing on years of experience in cross-community solidarity building and leading workshops on advocacy and anti-racism, she seeks to expand her practices for transformation through art and storytelling, using creativity as a tool to reimagine and embody new futures.

Mai-Anh Vũ Peterson (she/her) is a British Vietnamese strategist, writer, facilitator, community organiser and co-founder of the group Britain's East and South East Asian Network (besea.n). She leads accessible and inclusive workshops, rooted in justice, care and creativity, on topics such as grief, burnout and identity-based harm. She writes about the arts, identity, activism and community building, and is currently based in Edinburgh.

ACKNOWLEDGEMENTS

This book is a testament to the many hands, voices and histories that have shaped and guided us. It exists because of the people who came before us, those beside us, and those who will come after.

To our **ancestors**, whose strength, sacrifice and stories continue to shape us — we honour you.

To the **activists and changemakers, past and present**, who have paved the way, challenged the status quo and refused to let silence settle – we stand on your shoulders.

To **Delayna Spencer** and **Emma Yuan**, thank you for your belief, trust and patience in us despite the hurdles. We extend our gratitude to everyone at Sage involved in the production of this important series. To **Meredith Clark**, we thank you for your critical voice in helping us frame our writing.

To **Suyin Haynes**, for bookending our journey of this undertaking with much valuable insight, support and generosity.

To the **besea.n core team and wider network**, for your unwavering support, advice, love and acts of service – this work is stronger because of you. We thank all besea.n members, past and present, for their early work in mapping out our ideas. To **everyone we interviewed** for this book, you know who you are. Thank you for sharing your worlds with us: your insights gave us a rich picture of East and Southeast Asian (ESEA) lives, even if your words are not directly quoted.

To **Liz Pemberton**, for championing us.

And finally, to our **partners, families and chosen families**, for your love, patience, ideas and steady presence. You remind us why this work matters.

INTRODUCTION

The unfolding events of recent years have made one thing clear: we are steadily descending into a tangle of interconnected crises. Economic instability is deepening as corporations exploit inflation, while wages stagnate and the cost-of-living crisis hits marginalised communities hardest. Meanwhile, since 2010, the UK government has prioritised austerity over public investment (Blyth, 2013). The climate crisis – fuelled by racial capitalism and fossil fuel interests – is accelerating (Gonzalez, 2021), while imperialism and militarism drives global exploitation, enriching arms dealers and displacing millions (Barkawi, 2006). Instead of using technology to reduce our need to work, tech giants tighten their grip and people find themselves labouring more for less.

Social unrest is growing as groups of people resist exploitation, repression and systemic injustice. From labour strikes to environmental activism, grassroots movements worldwide are challenging the status quo – yet authoritarian governments increasingly appear to respond with crackdowns rather than reforms (Lakhani et al., 2023).

In the UK, East and Southeast Asian (ESEA) people make up a significant portion of the population[1], but have historically been perceived as less active in resistance efforts. In 2020, the COVID-19 pandemic catapulted anyone racialised as Chinese

---

[1]The 2021 Census of England and Wales showed a Chinese and 'Other Asian' population of 2.3%, but this does not account for Scotland, Northern Ireland, or mixed ESEA people in England and Wales.

into the spotlight, lifting the lid on years of entrenched racism and social exclusion (Gram & Mau, 2024). During this time, new organisations and groups were born, such as our own, besea.n (Britain's East and South East Asian Network), and existing ones expanded their remit to consider broader, more diverse understandings of what it means to be ESEA in the UK, such as the ESEA Community Centre Luncheon Club, which serves up a variety of weekly lunch dishes for members of its ESEA community (formerly Hackney Chinese Community Services Luncheon Club).

This book is an effort to reflect on the small space that we, the co-authors, occupy in resisting the systems that create these crises, through our experiences organising both together and in individual capacities. It's also an attempt to connect a few dots between the presence of ESEA communities, their experiences and their engagement in movements for social change. We have aimed to accommodate multiple reading styles, by writing in a way that allows for a non-linear reading of the chapters (see Chapter 7: Queering the Script for an explanation of Queer Temporality, and why we feel this is important). It also feels important to recognise the different struggles that have shaped the book-writing process alone: as a group, we've had to work through various forms of loss, a complete shift in ideas and approaches, a rapidly changing political world and the constant presence of self-doubt. Simply being charged with writing a book is complicated in itself; it suggests authority, expertise, prescriptive solutions. The ideas contained within its pages are a product not only of the individual experiences we've had and the activators, writers and thinkers who have shaped us theoretically, but of the communities and individuals we have organised with, who have shaped our practice. Ideas alone are not a solution for the crises we face, but writing this book has helped us to make sense of our own tumultuous worlds, and offered us a respite from the messy struggles of community organising.

A NOTE ON OWNERSHIP

This book was written in the spirit of the way we have organised together for the past five years: collectively, somewhat chaotically, and dynamically. We planned out our topics in meticulous detail together, with the idea that no single person would 'own' individual chapters. While we believe that people should be credited for their work (indeed, the works cited list is gargantuan), we want to challenge the notion of ownership of ideas within our own work. This is especially true since the attitudes and approaches we embody have been formed by so many different people, places and practices that it's difficult to identify who, what and where.

The sentences in this book have been touched by all of us, the product of a complex combination of in-person workshops, video calls, shared online documents and two years' worth of messages and voice notes. It's possible that this approach was more difficult and more time consuming than if we were to have divided up the chapters, and we'd be remiss if we didn't mention how much we struggled with it, at times wondering how we'd ever manage to produce something coherent. But moving in collaboration also offered us strength during times of increasing darkness. Activist and grassroots organising spaces are rife with friction, and our writing project gave us space away from the noise to reflect on the part we want to play.

THE BIRTH OF BESEA.N

In the summer of 2020, six strangers connected online to pool resources in response to the increasing racism, violence, hostility and fear directed at ESEA individuals and groups during the pandemic (Gram & Mau, 2024). They channelled their efforts into addressing the media preoccupation with plastering COVID-19 articles with stock photos of miscellaneous ESEA people, constituting a level of mainstream representation never seen before. Appealing to white journalists at huge media institutions bore no fruit except to grow their frustrations and call more people to the issues faced by ESEA people. The realisation that a large number of ESEA people lacked a suitable space to take ownership of their stories led to the development of the platform, besea.n.

Through the evolution of our group since then, we have endeavoured to create a place for ESEA people to explore a variety of issues relating specifically to ESEA experiences in the UK. We discovered and met many groups that had organised around a similar premise and our network grew as a result. Despite our core team organising across vast distances (at the time: London, Manchester, Glasgow and Senegal), strong personal connection in our early stages allowed us to safely build the foundations of friendship and reformulated our relationships with our own ESEA identities. Only after doing so were we able to focus on community action.

Over the years, our membership has changed, welcoming new changemakers and bidding farewell to others. In July 2024, we were devastated to the core by the loss of our dear friend and member, Peiwen Tian. The impact of her passing has profoundly shaped the book you hold in your hands, and ultimately, the future of our group, which finds itself in a new landscape of strategising, rebuilding and regrowth. While the book is a product of collective experiences as a group, the opinions expressed within are our own, the co-authors. We don't speak for our members, nor for the communities we work with, although we've tried to include as many insights and voices as we can, to illustrate the complexity of ESEA presence and activity on this small island that connects us.

ESEA HERITAGE MONTH

We find that taking up space in a society that doesn't always make us feel included is best done through creating opportunities for meaningful community organising. That's why besea.n founded ESEA Heritage Month in 2021, which takes place every month in September. What drives ESEA Heritage Month, however, is the unswerving passion and dedication of the event holders and advocates for this cause, who volunteer their time, and sometimes money, to hold space for critical moments of connection and education.

We are wary of the ways in which heritage months and heritage interpretation have become absorbed by corporate Diversity, Equity and Inclusion (DEI) practices that don't address meaningful change. For us, ESEA Heritage Month is a container of opportunities for us to better understand ourselves by witnessing stories and experiences that reflect realities similar to our own. The month serves as a thread to connect us, turning our geographic distance towards a fertile ground, ready for an exploration of individual and collective identity alongside greater empathy building.

We approach ESEA Heritage Month with a community first mindset: it is 'organised' by us in so much as we provide the platform and the network through which to distribute events and information, but the programme is shaped and led by individuals and organisations running their own events up and down the UK. It is crucial that this annual space for understanding – which, in a perfect world, would take up space all year round – exists beyond the individuals involved in its conceptualisation.

BEYOND BELONGING

Much of heritage interpretation and identity discourse in the last few years has focused on the concept of belonging – on feeling comfortable in one's own skin and acceptance within society (Hirsch, 2018). Early visions of this book centred around ideas of finding *our* place, as though 'place' was a static concept rooted in finality. It quickly became clear that we did not want to occupy a static place in a system that values profit over people, and shows little signs of meaningful material change for those at the sharp end of multiple global crises. Nor did we feel comfortable with aspirations of assimilation within structures that remain inherently patriarchal, heteronormative and eurocentric. At the same time, we also saw increasing celebrations of 'visibility' and ESEA representation – acknowledgments that often felt disconnected from the deep-rooted inequalities and harms unfolding in the UK and beyond.

And so through this book, we ask: what does it mean to go beyond belonging? How can we harness the connections over shared identities, histories and presence in this country to move through action and activation to a better system? How do we use intercommunity connections to embody the world we want to see? Perhaps the real question isn't whether we have the answers, but how we keep asking the questions that matter.

1

ESEA PEOPLE IN THE UK: A COMPLEX NARRATIVE

WHY NOW?

Type 'Asian activists' or 'Asian anti-racism' into a search engine and a plethora of results will link to articles written by, about, and for Asian Americans[1]. By contrast, in the early stages of researching this book, we ran into difficulties finding studies on East and Southeast Asian (ESEA) communities in the UK that weren't a footnote or a one line addition to a larger piece about South Asians or 'BAME' (Black, Asian, Minority Ethnic) communities[2]. Against a tide of rising anti-immigrant senti- ment in this country (Hackney Chinese Community Services,

---

[1]We intentionally write Asian American in this scenario rather than the more widely encouraged term Asian American Pacific Islander (AAPI) because we recognise the erasure of Pacific Islanders that takes place in activism that purports to be for AAPIs.

[2]'BAME' was used as a literature search term; as co-authors, we prefer to use the term 'Global Majority'.

2023), it feels more important than ever to understand ESEA presence, identities and struggles in a UK-specific context.

At the time of writing, the definition and experience of what it means to be 'British' is expanding and evolving to embrace more variance and possibility than ever before. Within 'identities' like 'British Asian', we may find a multitude of others: British Indian, Welsh Pakistani, British ESEA, Scottish Hakka, Hong Kong Londoner, British Viet Hoa, to name a few. Identity markers are informed by ethnicity, race, nation, city of residence, religion and many more. It is a beautiful thing not to be limited by semantics. Many ESEA people residing in the UK – even those with citizenship – don't use terms like 'English', 'Scottish' or even 'British' to describe themselves, yet their experiences are still very much part of the fabric of life in the UK (Southeast and East Asian Centre, 2023).

Over the last seven decades, as a result of the changing tides of migration, post-war economic shifts and evolving social behaviours, multiethnic communities in Britain have become more present in its society (Royal Geographical Society, 2023). With that, we have had to reckon with what it means to be a new kind of British. We have seen governments push to incorporate 'British values' into the teachings of schools and institutions, but those British values speak to a white, middle class experience many other communities in the UK cannot relate to (Vincent, 2019).

Recent years have seen global protests against the repeated killings of Black people by police in the United States (Amnesty International, 2021), increased scrutiny of the legacies of Empire, and more focus on the impact of systemic racism (Human Rights Watch, 2024). These have all meant that the

UK *seems* to have a much more enlightened racial discourse than it did twenty years ago. In terms of cultural shifts, the celebration of holidays like Diwali, Eid and Lunar New Year are now more commonplace in the public sphere. There are more initiatives dedicated to preserving the history and heritage of Black and Asian communities, and being against the 'mixing of races' is now no longer a majority opinion. One study suggested 89% of Brits would be happy for their child to marry into a different ethnic group (Ipsos, 2020).

While we certainly haven't achieved a 'post-racial' society or a liberated one, it's safe to say that things are very different now than after the post-World War II migration boom. It's important we remember this. Despite disagreements in political movements, it's undeniable that some things *have* changed. There are plenty of challenges that come with community organising, organic growth as ideas change and politics shift, but it's important we don't throw the baby out with the bathwater when we're considering what's working, what isn't, and what needs to change.

How do ESEA people fit into the history of civil rights movements and resistance in the UK?

Alongside the history of the Bristol Bus Boycott in 1963 by striking Black workers, and the collective action of young South Asians in the Asian Youth Movements in the 1970s–1980s, we should also learn of the striking Chinese seamen in Liverpool in 1942, or the mass mobilisation of ESEA workers in London in protest of racist media scaremongering surrounding the foot and mouth disease epidemic in the early 2000s. By calling attention to stories of historical resistance by a multitude of marginalised groups against similar struggles, we contribute to a practice in which systems of oppression are tackled collectively, allowing us to work in cross-cultural solidarity *with* each other rather than against each other.

LANGUAGE MATTERS

Every book on race, ethnicity and racial justice we've ever read addresses the semantic quandary of talking about racial identity. People of colour? Global Majority? Ethnic minority? Non-white (eek)? Racialised people? An article from the Public Data Lab, an interdisciplinary network exploring how digital spaces are used for social impact, highlights the double-edged sword of such terminology, which can both progress and limit the groups they refer to. While essential for mapping communities and ensuring institutional inclusion, they also risk becoming 'instruments of exclusion – particularly if they mis-represent the groups they're supposed to encompass. Indeed, communities often also form in resistance to the very mechanisms of categorisation' (Public Data Lab, 2021).

Perhaps nowhere is the diasporic fight to claim language stronger than in the USA. Although the term 'Asian American' emerged in the 1960s, it gained widespread use as a political identity after a wave of discriminatory acts that took place in the US, including the murder of Vincent Chin by two white Americans in 1982 (Asia Society, 2017). Protests arose due to the lenient sentences faced by the two men: a $3000 fine and no prison time. Chin, who was Chinese American, was racialised as Japanese by the perpetrators, who were seeking to lay the blame for the declining American auto industry at the feet of Japanese auto companies. To many Asian Americans, this exposed the precarious common ground on which they stood – that they could face violence as Asians regardless of their ethnicity.

Institutionally, the acronym 'AAPI' (Asian American Pacific Islander) came into use, though in time, its ineffectiveness to represent such a large group was laid bare. Pacific Islanders, Southeast Asians, and South Asians often feel invisible and at odds with the standard perception of AAPI people as East Asian, while also having their specific struggles erased. The AAPI category also obscured existing inequities, 'making it seem as though Native Hawaiian and Pacific Islanders were overrepresented at higher education institutions like other Asian Americans when the opposite was the case' (Zhou, 2021).

We would see this same kind of non-specific, blanket racism emerge with the spread of the global COVID-19 pandemic. During this time, increasing use of the term 'ESEA' acted as a banner, under which many who were racialised as Chinese faced increased discrimination and violent racism (University of Leicester, 2024), as they became synonymous with the virus itself. The highly publicised and unprovoked attack on Singaporean Jonathan Mok in London in 2021 – in which the perpetrators shouted, 'We don't want your coronavirus in our country' – was a wake up call for many ESEA people. Mok survived the attack, but required extensive facial reconstruction surgery (Ng, 2021). ESEA communities were already feeling targeted by the reporting of mainstream media and dog whistles from politicians, such as Donald Trump, weaponising terms like 'China Virus' and 'Kung Flu' (*The Guardian*, 2020).

The subsequent political organising that emerged can be viewed similarly to the mobilisations following the Chin murder in the US, a country that has a richer history of debate

and engagement around political racial identities. Activists
and researchers like Dr. Diana Yeh and Jonathan W.Y. Gray
have been instrumental in early research on the emergence
of the term 'ESEA' as a political, as opposed to a cultural or
linguistic, formation. There is the indication that communities engaged in this movement in the early 2020s found
themselves in a stage of 'becoming' ESEA, implying the
autonomy and intentionality behind embracing political clas-
sifiers, as we have seen with 'Asian American' and its varying
forms (Yeh, 2021).

CLASS AND THE MODEL MINORITY MYTH

The model minority myth is a stereotype that emerged in the
United States during the mid-20th century, primarily targeting Asian Americans, but it is one that has been extended as
a common belief with regards to Asian people in diaspora all
over the world. The myth suggests that Asian people have
achieved higher levels of success than other ethnic groups,
through hard work, strong family values and cultural traits,
rendering them more valuable to society (Petersen, 1966).

However, Asian people belong to hundreds of different ethnic groups and speak more than 2,000 languages,
so it's absurd to assume they all exhibit the same attitudes
and behaviours. The myth doesn't take into account important intersections such as economic or migration status, nor
the colourism that Brown South and Southeast Asians may
encounter. By perpetuating the model minority myth, society
tends to ignore systemic factors such as racial discrimination,
inequality, economic systems and historical disadvantages that
contribute to disparities among different marginalised groups

(Ruiz & Tian, 2023). By reinforcing the idea of meritocracy during a time of heightened anti-immigrant sentiment, the model minority myth creates dangerous standards around who should be shown humanity and who shouldn't.

The myth can also create unrealistic expectations for Asian people across different ethnic groups, placing immense pressure on them to conform to high standards of achievement. It can lead to feelings of isolation, mental health issues and internalised racism when individuals are unable to meet these expectations, or when their experiences do not align with the stereotype. We see examples of how this impacts ESEA people in our education system: East Asian students are often perceived as hard-working and lacking in critical thinking (Moosavi, 2020), and a study from the University of Bristol suggested that ESEA students are typically over-assessed by teachers, in comparison to Black students, who are under-assessed (Burgess & Greaves, 2013). Subsequently, ESEA students are often held up as examples to prove that ethnic minority students are not disadvantaged in schools – essentially, a denial of institutional racism. Furthermore, this myth means that these students' needs for pedagogical and mental health support are often overlooked as staff don't see potential issues due to the stereotypes that ESEA students thrive academically.

It's also critical that we take economic status into account when we consider how people from different ethnic backgrounds experience society and access services. An ESEA kid growing up in a low-income household, on free school meals, is likely to have shared experiences with other people in similar economic backgrounds from all ethnicities. They may have much less in common with an ESEA kid who has wealthy, upper middle class parents, who attends a top private school.

Furthermore, the way we identify racism in society depends on class, too. Writer and artist Akala has observed that we have been trained to denounce racism committed by poorer people, characterised through stereotypes of EDF-supporting hooligans, while turning a blind eye to the enduring atrocities of colonialism, which have historically been led and perpetuated by wealthy elites. 'This neat binary,' he says, 'is a great way of avoiding any real discussion at all' (Akala, 2018).

The model minority myth showed up very clearly in 2021 in a government report by the Commission on Race and Ethnic Disparities, known commonly as The Sewell Report. Its aim was to investigate public evidence and determine the extent of institutional racism in the UK, and was chaired by Tony Sewell and commissioned by Munira Mirza, who have Black and South Asian heritage respectively. Both had expressed scepticism around institutional inequality in the past, with Mirza going as far as to write about the 'myth of institutional racism' (Mirza, 2017). Is it any surprise, then, that the Commission came to the conclusion that the UK is not institutionally racist? The report, which was widely discredited by experts in academic and activist spaces, neatly summarised the following: 'Education is the single most emphatic success story of the British ethnic minority experience' (Commission on Race and Ethnic Disparities, 2021).

Voilà. The model minority myth rears its ugly head again. If ethnic minority groups are considered successful for performing well in school, what does it say about ethnic minority groups who do not perform well? And what happens when we combine this with stereotypes that Asian people are hardworking and quiet?

The erasure of racist experiences endured by ESEA people is further compounded by the fact that the report relied on ethnicity categories from the England and Wales Census. Beyond Indian, Pakistani, Bangladeshi or Chinese groups, the report gave no specific findings for any other ethnic groups.

WHO GETS TO BE ASIAN?

What do we think of when we hear the term 'British Asian' in the UK?

Most people would likely imagine someone of South Asian heritage. According to the 2021 Census of England and Wales, Chinese and Other Asian ethnic groups were about a third as big as the combined number of Indian, Pakistani and Bangladeshi groups. However, an undefined margin of error must be taken into account when considering population demographics, due to insufficient disaggregated data. This has an impact on the way people perceive 'Asians' typically. A study on Asian subgroups in the United States and the United Kingdom found that Americans tend to think of ESEA groups when they think of Asian people, while Brits are likely to think of South Asian groups instead (Goh & McCue, 2021).

The way in which UK society views the term 'Asian' is reflected in the target audience for the publicly-funded BBC Asian Network channel. The description on its Instagram page states its purpose as 'Celebrating British Asian identity'. However, anyone expecting to discover the latest in Cambodian rap or Malaysian pop will be sadly mistaken; it's clear that the network primarily features artists of South Asian descent.

Many ESEA people who feel overlooked by institutions that use the term 'Asian' to mean 'South Asian' advocate for a more nuanced approach to language, preferring specificity rather than terminology that risks misrepresentation. Writer Nikesh Shukla (N. Shukla, personal communication, 19 June 2023) has often referred to the channel as the 'BBC South Asian Network', mentioning the 'political necessity' of such a transition because of what he calls a 'flattening' of diverse communities. Shukla also highlighted BBC Asian Network's 'disconnect from Asian culture,' reinforcing the erasure that many Asian people feel, which causes a mutual alienation between these institutions and the cultures they claim to represent (Aujla-Sidhu, 2021).

Then there's the fact that many ESEA people are assumed to be Chinese. Dr. Diana Yeh spoke to us of 'Chinese-ness [as] the default Asian identity for East and Southeast Asians in the UK' (D. Yeh, personal communication, 9 August 2023). Throughout her work, Yeh has argued that 'Chinese' has become a hegemonic racial category in the UK, erasing and subsuming other East and Southeast Asian identities by functioning as a racial rather than ethnic label rooted in colonial legacies and racial hierarchies. The colonial history of Hong Kong as a British colony, as well as a more informal imperial relationship through Chinese treaty ports, strongly influenced migration patterns to Britain and perceptions of people from the ESEA region (Yeh, 2018; 2021). The blurring of ethnic lines that occurs when people misrecognise all ESEA faces as Chinese means that specific types of racism such as sinophobia (anti-Chinese sentiment) impact many non-Chinese ESEA people.

Additionally, according to Yeh, the fact that China and Chinese-ness have a dominant position within the British imagination brings fixed notions of what it means to be 'authentically' Chinese, which obscures the diverse experiences of many different Chinese identities (D. Yeh, personal communication,

9 August 2023). How can there be an 'authentic' way to perform a hegemonic identity that does not really exist?

The systemic erasure of ESEA people by public institutions and the lack of disaggregated data can lead to their being absent from conversations around race. As mentioned, the commission responsible for the Sewell Report lacked ESEA representatives, as did the panel which the Runnymede Trust (2021) convened in order to question its findings. We remain sceptical of whether increased data collection among marginalised ethnic communities is always beneficial (see Chapter 3: ESEA people in the UK today). However, we should acknowledge that such erasure limits the opportunity for ESEA people to move collectively against issues that impact multiple communities. Exploring this struggle within a community setting would allow them to express their needs for culturally aware supportive services or informal spaces.

ESEA-focused community centres – such as ESEA Community Centre (formerly Hackney Chinese Community Centre), SEEAC (Southeast and East Asian Centre) and the Filipino Women's Association – carry out important work to tackle the exclusion felt by ESEA people in British society. Often, these centres and organisations are severely underfunded and rely on a small number of financial benefactors or arduous grant writing applications, which we have seen lead to burnout for their employees and volunteers. The critical support that the most vulnerable ESEA people – such as undocumented migrants, domestic workers and the elderly – rely on is more likely to exist under precarious circumstances because of this underfunding.

The data gap for ESEA communities furthers their invisibility in public discourse. The knock-on effect of this cycle reduces organisations' access to funding and, by extension, their ability to carry out community building and outreach for people who need it the most, thereby further reducing public and institutional consideration in a vicious cycle. This highlights the unsustainability of current funding models and serves as a reminder that the NGO industrial complex (see Chapter 8: When change meets resistance) is deeply flawed. However, the more urgent issue remains: essential services are abruptly shutting down, with few alternatives available.

A PARTICULAR BRAND OF RACISM

The racism towards ESEA people, and Asians more generally, has always emphasised perceptions of foreignness, alienness and savagery. Readers, we must admit that when we were writing this section, we did wonder if it would read as a 'How To' guide for racism against Asian people. But some things need to be plainly stated.

While some slurs have faded with time, others are still in use today. It is still perfectly acceptable to many people in the UK to refer to their local Chinese takeaway as 'the Chinky'. Eye pulling is a regular occurrence, one that follows you from the playground to the workplace. ESEA people are also frequently assumed not to speak English, while ESEA languages are often reduced in mimicry to meaningless noises intended to sound vaguely Chinese. ESEA foods are considered weird, smelly, and there is a popular perception that ESEA people will eat anything that walks, crawls or swims.

In addition to stereotypes informed by the model minority myth – such as submissiveness and meekness (Yoo et al., 2010) (which arguably make ESEA people likelier targets for violent racism) – they also suffer from what we might call classic racism, which most other racialised people are sadly all too familiar with. This might include, but not be limited to:

People confusing two ESEA people who look nothing alike

People refusing to learn how to pronounce an ESEA person's name

Perceptions that ESEA culinary habits are barbaric and unhygienic

People expressing disapproval about how many ESEA people there are everywhere (an influx, a horde, a swarm)

Local businesses being vulnerable to harassment and vandalism

Then there's Southeast Asia specifically. Against the trope of the East Asian concubine is an imagining, shaped by a Western gaze, that women of Southeast Asia are synonymous with sex work[3] (Espiritu, 1999). Southeast Asia, and Thailand in particular, are associated with cheap sex and sleaze (Chee, 2015). But the same salacious lens is not applied to anywhere else in the world with a busy sex work industry. In fact, using UNAIDS data from 2016 on sex worker populations and overall country population, Wikipedia contributors created a list of 'prostitution prevalence', ranking countries by number of sex workers per 10,000 population using data from UNAIDS. There are no Asian countries in the top 20, let alone the top 10. Thailand

---

[3]While we as co-authors maintain a sex positive approach to sex worker rights, we acknowledge here that sex work is still seen negatively among many in society.

ranks at number 66, far outstripped by the USA, Germany, Austria and Croatia – Taiwan comes in at 24 (Wikipedia Contributors, 2019).

In diaspora, particularly in the Global North, ESEA people tend to be considered with their Asianness centred, but often simplified or misrepresented (Lin, 2022). This kind of ESEA essentialism is tricky to navigate, because the very things that people don't like to have imposed on them are often things they themselves want to embrace and assume. For example, a British Korean writer may wish to centre Korean cultural traditions and East Asian influences in their work, and be recognised professionally for the role their heritage has played in informing their perspectives. However, they may not wish to be 'pigeon-holed' as 'the Asian writer', responsible for covering all things Asian, even subjects in which they're not remotely expert.

Identity-driven exploration and creativity, whether personal or professional, can be incredibly rich and rewarding, but the fact of the matter remains: being Asian is not one's whole identity.

SINOPHOBIA

As mentioned earlier, sinophobia has shaped much of the ways in which ESEA people are treated and the extent to which they are accepted in diaspora around the world (Tchen, 2005). Sinophobic attitudes tend to conflate ethnicity and nationality, informed by stereotypes associated with the Chinese state. An example of this occurred in 2022, when Labour MP Neil Coyle was suspended from the party for making a joke about a British Chinese journalist being a Chinese spy (*BBC News*, 2022). Historically, sinophobia has manifested in migration policy in the USA, through anti-communist sentiment since the Cold War, US-China geopolitical tensions and media representation (Ngai, 2004). It has naturally found its way into the social imagination through various cultural and racial stereotypes, which can impact anyone racialised as Chinese.

In popular culture, sinophobia has also taken on a particular aesthetic. Walk into any bookshop section on East Asian history, politics or economics, and the dominant, angry, red book cover that symbolises 'China and All Of Asia' is hard to miss. Maybe the illustrators will throw in a dragon, some stars, or a rising sun for some variety. This flattening of Chineseness is also found in perceptions worldwide about Chinese products: they are perceived as cheap, and therefore of low value, tacky or even dirty. This devaluation extends to the bodies who produce the goods, meaning that Chinese labour is also universally considered as cheap (Pai, 2008). It doesn't even matter that 'Chineseness' within China itself isn't an essentialist notion, with more than 56 ethnic groups recognised in the country, let alone the diverse spread of Chinese diasporic, transnational groups outside of China.

Racism knows no borders.

There is a particular attention paid to China when it comes to moral policing and political commentary. It's a centuries old suspicion that hasn't really gone anywhere; it's simply been upgraded with some 21st century flair. A common perception of China these days is that it is a new imperial power, intent on global domination. Its human rights abuses in Xinjiang have been widely condemned as a form of colonialism and ethnic cleansing, and its territorial disputes in the South China Sea reveal an obvious expansionist project (Center for Preventive Action, 2024). However, the same level of criticism is not levied at the USA, which has over 750 military bases worldwide and is involved in active, imperialist wars to this present day (Haddad & Hussein, 2021).

For the avoidance of doubt, this is not about weighing up US versus Chinese imperialism. But it is important to highlight that anti-Chinese sentiment is heavily influenced by a western perception that West = good and China = bad. This impacts the racism that ESEA people experience, whether they have anything to do with China or not. Suggestions of Chinese imperialism are also often used to justify US military presence, influence and even control in the Global South, with the Philippines being a good example (McCoy, 2009). While Filipinos themselves are divided over the issue, the question remains over whether the threat of one imperialist power is worth sacrificing autonomy at the hands of another.

A RICHER SOCIETY

There is no doubt that we live in a country that has become richer and more diverse even over the course of our (the co-authors') lifetimes. One need only marvel at the sheer range of shops, restaurants and services available in cities like London, Manchester, Birmingham and Glasgow. From pandan flavours to batik textiles, ESEA influence on the tastes and trends of UK consumer sectors is growing in strength. Migration links with its formerly colonised regions mean that the UK can be considered to have some of the best Malaysian and Cantonese food outside of Asia, and no one can deny the explosion of Taiwanese-inspired bubble tea shops up and down UK high streets and in shopping malls.

While bubble tea might seem like a superficial addition to a consideration of migratory societal impact, what it represents is a willingness to embrace flavours and textures previously unknown to the 'typical' western palate. We see the same acceptance with the popularity of mochi ice cream such as Little Moons Mochi, which has taken the UK by storm (Woolfson, 2021). As the demographics of our society change, so do the demographics of our workforce. What studies generally show is that increased net migration has had a positive impact on the British economy (UCL, 2014).

We should stress that we don't believe that attitudes to migration should be informed by how much it contributes to society. Doing so risks leaning into hierarchical racial stereotypes about good immigrants and bad immigrants, and deciding whether or not people should be afforded basic rights based on their economic contributions and outputs. However, in a political and media landscape that is increasingly hostile to migration and considers it a drain on public finances (Carlile & Harrison, 2022), it feels important to mention that London's 1.8 million migrant workers contributed around £83bn to the capital's economy in 2017 (UK Data Service, 2017).

A report from 2019 found that one in ten Scottish businesses were led by an immigrant entrepreneur, contributing more than £13bn to the Scottish economy and creating over 100,000 jobs (The Federation of Small Businesses, 2019). For Chinese communities alone, the annual Tou Ying report – which measures UK Chinese business activities – showed that in 2020, UK Chinese businesses had a combined turnover of £92bn, creating more than 75,000 jobs nationwide despite pandemic disruption.

As migration patterns shift, so too do the demarcations of regional hubs like Chinatowns in the UK. In London, for example, food writer Angela Hui argues that the capital no longer has just one Chinatown – it has many (Hui, 2020). Furthermore, Chinatown is no longer made up solely of Chinese businesses, but a panoply of ESEA offerings catering to a widely changing migrant demographic. Other blossoming enclaves of ESEA migrant communities of the last few decades include Chinese Uyghur populations in Holborn, Vietnamese in Hackney and Lewisham, and London's Koreatown in New Malden.

As communities have grown and transformed, the last few years have seen something of a shift in terms of cultural outputs centring and celebrating ESEA experiences. An undeniable cultural phenomenon of recent times is the global popularity of South Korea's cultural exports including pop culture, fashion, cosmetics, entertainment, music, TV dramas and movies. This phenomenon, known as Hallyu – a Chinese term which, when translated, literally means 'Korean Wave' – was the topic of a special exhibition at the UK's V&A Museum from September 2022 to June 2023 (Adams, 2022).

In literature, our shelves have been graced with more covers from ESEA authors, no doubt supported by the efforts of publishing leaders and community organisers within the literary sector like the group ESEA Publishing Network. Elsewhere, we see performers in the music, TV and theatre sectors collaborating to address injustices and systemic exclusion on our stages and screens, and behind the scenes. Industry-specific groups like ESEA Music and MilkTea (a UK-based ESEA collective showcasing ESEA cinema and organising community screenings) have sprung up in the last few years, joining the longstanding creative and activism efforts of groups like BEATS and Kakilang (formerly Chinese Arts Now).

In short, ESEA communities are more visible in public consciousness than they were even five or ten years ago.

SO WHAT NOW?

The rest of this book will – through collective writing, conversations between ourselves and insights from people we have interviewed – attempt to provide an understanding of how ESEA community organising has evolved. In order to understand where we're going, we need to understand our past and our present, the achievements and the challenges. Like any political movement, a healthy amount of disagreement and debate is perfectly normal. Organisers may agree on the end goal, but be at odds with each other over how to get there.

Increasingly, we've seen the harm produced within our own organising circles by all too familiar patterns: physical and emotional burnout, conflict, rigidity. 'Growth' is a word we see used a lot, yet most of us are yet to develop the reflex to make space for it. Even writing this book is a case in point: a collective writing process that started in 2023, it has taken many forms as its authors' ideas, experiences and values have shifted and transmuted. Perhaps we'll look back at it in ten years when those ideas have shifted yet again and, we hope, accept the people we were, rather than cringe at the journey that got us here.

As we have mentioned, ESEA narratives are more visible than they were even just a few years ago. But those shifts alone have not caused any meaningful change to the material conditions of the most vulnerable people in our society, who are at the sharp end of systems of harm. We must redirect our focus to creating spaces that move in solidarity with those people, reduce those harms and offer respite, rather than expecting trickle-down change or widespread policy change. Of course, there are plenty of things to be glad about: it matters to be able to see a wide range of experiences represented in the media we consume; it matters that it's no longer acceptable to use slurs in the workplace or denigrate people's food; and it matters that ESEA people feel more emboldened to advocate for themselves with agency. But how do we move beyond the needs of the individual?

We have dedicated an entire chapter at the end of this book to imagining the future of activism, with an approach that is grounded in Black feminist, anti-capitalist traditions. We believe that the work we have ahead of us involves developing and honing what adrienne maree brown refers to as 'the practice ground for what we are healing towards, co-creating.' It's critical that we remember that these movements and networks of care are often being carried by people who are themselves deeply wounded and vulnerable, but who are also products of a broken system. Or as brown puts it: 'We need the people within our movements, all socialized [sic] into and by unjust systems, to be on liberators' paths. Not already free, but practicing [sic] freedom every day' (brown, 2020).

Building new reflexes into our work is hard, and brown is correct: we need all the practice we can get. We also need grace, which allows us the space to learn from mistakes, and afford ourselves care and rest when the familiar patterns of harm start to creep in. Mobilising through care allows us to reimagine systems of working that lie beyond the rigid boundaries of our current systems. At besea.n, at the time of writing, we are working through ideas of seasonal or slow organising, collective tasking, and imagining activities that reduce reliance on funding. All of this is in the service of care not only for ourselves but our communities. This tiny sliver of a new horizon has allowed us to begin to imagine what it could look like if care-centric organising methods and interdependencies were practised not just at grassroots level but embedded into institutions and economic systems.

The landscape we operate in is different to what it once was, and the speed at which the digital world is transforming suggests that change will be the only constant, especially as the world continues to deal with staggering levels of inequality. Advocacy, activism and community organisations have existed for specific communities since the first ESEA groups migrated to our shores, and we must learn from the long-term work of these pioneering organisers. Although anyone can be an activist, something we have felt the need for in our work is more mentorship and knowledge exchange that fosters interdependent networks of care and development. So many people are thrown into organising via a baptism of fire and must learn how to safeguard themselves and others the hard way. Events like teach-ins offer a more participatory style of knowledge exchange and awareness raising but, unless facilitated well, these too can risk being inaccessible for those outside the bubble.

It's important to safeguard the work that has been done so far in order to support the work of generations to come, including working actively in solidarity across generations to plan for the future. Younger generations have always been instrumental to the changing dynamics of community organising and networking, relative to the times. Second generation ESEA people are partly responsible for the greater spread of ESEA populations in the UK today, leaving the hometowns their parents originally settled in, migrating to new cities for work and study. These moves coincided with the rise of social media and a new digital age of rapid information sharing, allowing easier mobilisation for political issues.

As more third generation ESEA people grow up, the nature of community organising will shift again. A new age of reckoning with systemic oppression across lines of class, race and gender has taken hold in the UK, and we must take the necessary steps to collectively channel the rising tide of primarily digital ESEA community organising, ensuring a mobilisation that goes beyond 'belonging' and reaches for something like – could it be? – liberation.

2

IDENTITY IN MOTION

THE MOSAIC OF IDENTITY

This is not a book about 'the British ESEA identity', or even 'ESEA diaspora identity' in the UK. Even simply using the term 'British ESEA' suggests a singularity of experience, yet it encompasses a vast and often conflicting mosaic of individuals, cultures, religions and histories. From the child of Chinese takeaway owners in Birmingham to the Malay Muslim student in East London, ESEA experiences are shaped by migration, class, race, gender identity and a catalogue of other traits so vast it would put Argos to shame.

While some people embrace labels like 'British Asian' or 'British ESEA' for political reasons or even just for convenience, we found ourselves asking why it mattered to have ways of understanding and exploring changing identities, especially in community building work and efforts that strive for liberation from oppressive systems. While this book *does* explore the complexities of ESEA diaspora identities (plural intended) in the UK and how they are formed, challenged and redefined across generations, it's also about how, why and when we gather to create change.

The reflections that follow are shaped around an honest conversation between the original co-authors, in a video call, one Saturday morning in 2023. While we initially leaned on Kimberlé Crenshaw's framework of intersectionality for insights into ESEA-specific contexts such as British Chinese, and how our identities are shaped by the digital age, as well as tools of heritage interpretation, we have since looked to the work of academics Diana Yeh, David Parker, Miri Song and Yujie Zhu.

Language, 'class' (that nebulous term deeply embedded in British culture yet defined differently by everyone), financial wealth (or lack thereof) and education have all shaped our experiences of the world, which have continued to evolve as our identities have developed. Our conversation led us to understand how different and fluid our identities are, even for a group of people from a very similar demographic in terms of ethnicity, age and formal education. However, we also struggled with understanding our own experiences – as a group of relatively comfortable people with access to higher education – relative to wider material oppression of poor, more marginalised communities, asking ourselves how much we should focus on identity while wider issues of systemic racism and environmental destruction were tearing people's lives apart. While we dedicate another chapter to the ways in which digital spaces have shaped our tendency to think in terms that are mutually exclusive or polarised (see Chapter 8: From the streets to the screen), these are still questions that we as a group have asked ourselves. We have to reckon with the answers, if we truly want to dismantle systems of harm and embody the practices we wish to see in a better world.

ESTABLISHING NEW GROUND

To be a diasporic or transnational[1] person is to be considered Other[2], which is particularly true for Asian people, who are often described as being the 'perpetual foreigner' – a term coined in 1966 by William Petersen in an article about Japanese Americans (Petersen, 1966). Asian Americans are frequently judged to be non-American, despite their citizenship, accents or self identification. One study by the Pew Research Center found that 78% of Asian Americans have been treated as a foreigner at one point in their lives, either through being told to return to their home country, being assumed not to speak English or being criticised speaking other languages in public (Tian et al., 2023).

Asian people in the UK experience a similar perception of foreignness (Goh & McCue, 2021). Such perceptions have a likely impact on feelings of identity and self-worth, with one study finding that Asian Americans were more likely to use identity assertion techniques to demonstrate American-ness when they felt their identity was under question, such as knowledge of popular culture and denial of heritage (Cheryan & Monin, 2005).

---

[1]Diasporas are people with similar heritage or the same homeland who have moved out to different global locations. Transnationals are groups of people who maintain family and economic links across borders.

[2]'Other' is a philosophical concept that refers to individuals or groups who are perceived as fundamentally different or alien in relation to a dominant self or group. It can also be an action imposed on those people; the act of rendering them Other.

Studies show that there is a strong connection between chronic exposure to racism and discrimination, and both physical and mental health (Schouler-Ocak et al., 2021), but it is only recently that studies are starting to emerge to disclose the impact of secondary racial trauma. This occurs when individuals hear about or witness racism enacted against people of their own racial or ethnic background. A 2021 study on vicarious racism and the impact of constant vigilance in *anticipation* of racism was conducted on individuals of Black and Asian heritage in the USA in the wake of the COVID-19 pandemic. The results of the study found that increased vicarious racism and vigilance about racism led to higher rates of depression and anxiety among the impacted people, who were likely to derive self-understanding from values attached to their social groups (Chae et al., 2021). The study concluded that, with those groups under threat, so too are the self-esteem and wellbeing of the individuals within the groups.

Meanwhile, although we recognise the challenges associated with grouping communities and individuals together under catch-all terms like 'British Asian', 'trauma bonding' with other people from shared or similar backgrounds is common. We could even argue that a reclamation of shared identity in this way can act as a rallying call for political causes.

Professor Miri Song, whose work explores the relationship between ethnicity, identity and belonging in multicultural British contexts, suggests that, although self-designating terms like 'Black British' and 'Asian American' offer agency to individuals in curating identity, the extent to which people are able to negotiate and assert their ethnic identities depends on the social structures and constraints that surround them.

White people living in places like the USA and Britain can celebrate 'individualistic symbolic ethnic identities' such as Irish or German, which they can enjoy and invoke when they choose. By contrast, for people racialised as ethnic minorities living in white majority societies, exercising of 'ethnic options' as Song calls them, is often obscured or derailed by the fact that society imposes a *racial* identity on those people (Song, 2003).

However, Song argues that the assertion of ethnic identity is 'a central part of [ethnic minority groups] everyday lives' and is crucial in forming the way in which one responds to and perceives the social world they inhabit. This push–pull between imposed racial identity and personal ethnic identity is something that we have found among many in the ESEA communities we have worked with; a frustration at the societal assumption of an Asian monolith and the perpetual implications of the model minority myth.

What has become clear to us throughout our research for this book and the range of work we have done over the years is the significant gap in work dedicated to ESEA communities in Britain specifically. We have anchored our understanding in research like Diana Yeh's, unpacking the plurality of Chinese diasporic groups and pushing back against the 'hegemony' of Chineseness, alongside our experiences working with people from a multitude of different ESEA backgrounds. These have been important tools to help us move away from the tendencies of many in our circles (including us, at times!) to rely on the convenience of a singular experience: the collective, imagined 'we', the catch-all 'ESEA community'. Such a community, such a 'we', does not really exist, and an assumption that it does can exacerbate community tensions.

CYBERSPACES FOR HERITAGE EXPLORATION

Public availability of the Internet from the early 1990s meant the birth of a number of websites dedicated to social connection among different groups, such as British Chinese. Alongside David Parker, Miri Song explored this phenomenon with regards to the idea of social capital (the value or benefits that come from relationships and connections between people) and how it impacts participation in political pursuits. Parker and Song have pointed to Internet access as a critical part of exercising social capital and exploring ethnic identity (Parker & Song, 2007). This point is perhaps less obvious in the 2020s, in which digital access is much more widespread than when their 2007 study was published. However, we can still accept that having unlimited access to data and decent internet – both in terms of time and money – is still a dividing line. Those who work long hours in physical jobs have less time to engage with online advocacy and activism.

Parker and Song have also pointed out that, even in these earlier days of online political engagement, Internet advocacy and activist spaces were prone to fragmentation and commercialisation. It seems not much has changed in that respect. They point to these social networks – developed previously on websites and now, at the time of our writing, through social media and news sharing platforms – as being intrinsically tied to what they call 'second-generation' social capital. This is a product of collective awareness and networking, which 'generates dialogue, debate, and emotional support, as well as online and offline networks that facilitate social and political mobilization [sic]' (Parker & Song, 2006).

In contrast, those who engage primarily in online awareness campaigns may miss out on the benefits of in-person organising. Face-to-face interaction can foster stronger relationships and a deeper sense of empathy than communicating with a faceless mass of online followers (Turkle, 2015). We should not underestimate the benefits of a well-facilitated discussion without the distractions of social media feeds, where accommodations can be made for those traditionally excluded from online environments. With the appropriate care and invitations extended to impacted groups, attention can be focused on the matters at hand and lead to more productive outcomes. In person gatherings can also provide opportunities for clearer communication, where misunderstandings can be quickly addressed. The dominance of online conversations by those who can use fancy Canva templates and manipulate search engine optimisation (SEO) tactics is directly tied to a level of privileged access that excludes a vast number of people who do not have a proclivity for catchy infographic design.

We live in a time of distinct polarisation over identity politics that has been, in recent years, criticised on both sides of the political spectrum. On the left, many people believe that a focus on identity in political movements for change undermines traditional ideas of class solidarity and unity, or that identity politics are rooted too strongly in idealism, rather than in material realities. On the right, it's often dismissed as nonsense wokery. However, Parker and Song have argued that for groups like second generation British Chinese, building community foundations based on shared experiences and identity is a necessary part of collective mobilisation due to historical marginalisation:

> For the initial descendants of migrants, lacking a long tradition of political activity, the nurturing of bonding capital and particularized trust may be a necessary precursor for bridging activities, given the starting point of political marginalization [sic]. (Parker & Song, 2007)

We could also apply this logic to second and third generation ESEA people and some ESEA first generation migrants. Parker and Song cite criticisms of the social capital model, in particular of its treatment of 'bonding' practices, which are typically viewed as exclusionary, as opposed to 'bridging' practices, which tend to be seen as more effective between individuals with different interests and backgrounds. However, Parker and Song argue that the bonding/bridging binary fails to capture the importance of optimal use of *both* practices, particularly when it comes to ethnicity, and especially for second generation ethnic minority groups.

Certainly, at besea.n, most of us felt that a lack of community connectedness in childhood and young adulthood led us to feel starved of connection. In our early days of organising together, we bonded quickly and eagerly over shared experiences, both good and bad. We felt naturally inclined to bond over the interpersonal topics before transitioning towards mobilisation for broader and more serious causes. Indeed, many of us now view our early period of bonding as a necessary rite of passage.

CULTURE COLLIDING WITH (HEALTH)CARE

One topic that we returned to several times in our conversation was the idea that sharing similar experiences of ethnic identity with people you encounter in the world around you *can* help to ease navigation of public services such as healthcare, mental health journeys and care-focused practices. These services and practices may be culturally at odds with the traditions of the households some of us grew up in. Some of us raised experiences of being overlooked and dismissed in conversations with healthcare and mental health practitioners, questioning how much ideas like the model minority myth had pervaded those practices.

For example, one besea.n member was told by their (white) GP that they didn't need therapy, they simply needed to be more confident and assertive. Another GP reacted to an admission of severe mental health struggles, caused by racism, with sarcasm. Meanwhile, almost all of us felt that connecting with other people from similar ethnic backgrounds to discuss mental health was pivotal in providing supportive, reassuring environments for exploring and questioning the impact that identity had on wellbeing.

We spoke about the tendency for the children of immigrants and refugees in particular to have complicated feelings of survivor's guilt, deeply entwined with filial expectation to perform well in school and the professional world. This led to a discussion of how pain is recycled from the past and manifests itself in disruptive ways – a phenomenon felt across many different multicultural, multiethnic communities in Britain (Bhugra & Becker, 2005).

ASSIMILATING BETWEEN STANDARDS

As well as those with different passports, economic status, gender identity or sexuality, people living in bodies that don't fit British society's ideal standards of beauty often have to reckon with how their differences shape their identity. While this is true for people of all ethnicities, a particular focus in ESEA cultures on petiteness, especially for women, and specific ideas around how to perform femininity and masculinity were a recurring theme in our discussion. While one of us was taught as a young boy to take up space (quite literally, to manspread!), most of us socialised as girls were taught to be deferential.

The double bind of an open critique of physical characteristics by family members and constantly falling short of popular trends set by the white median was a formative part of our childhoods. After exchanging a few jokes about the various cringeworthy ways in which those of us who grew up in the UK tried to assimilate, our conversation turned to more serious revelations over negative self worth and low self esteem. After all, being young and self conscious is difficult enough, without being told that your nose is too flat, your eyes too narrow and your hair too dark. Many of our ESEA readers will also be familiar with the old 'why aren't you eating? / you're getting fat' dichotomy often levied at us by Asian mothers and aunties of a certain generation, and the pressure is especially acute for women and girls who are not married. Daughters and nieces finding a good husband and procreating as the end goal for many immigrant matriarchs was a common experience raised in our discussion.

As attitudes towards body image shift in a more positive direction in the West, those of us who grew up here often struggle with the return to our parents' countries of origin, where unsolicited comments about our bodies are as common as saying, 'how are you?'. While we can acknowledge that the scrutiny on body image can be interpreted as a form of care

in some ESEA countries and among elders of a certain generation, the comments may still sting.

Children of the first generation ESEA migrants who came to
the UK in the mid to late 20th century are now becoming parents themselves, filtering and interpreting the messages they
received from elder family members about what it means to
care and show love. Perhaps for those who undertake the work
of instilling a strong sense of cultural awareness and pride in
their children, there is also a path towards reconciliation with
past confusions and self doubt.

SPACE, PLACE AND COMMUNITY

For many people who grow up in diaspora or transnational
families, experience is shaped by a complex interplay of geography, culture and community. To be able to connect with
friends from shared backgrounds, explore and question one's
identity in a safe environment may even enhance other friendships and connections with others from different backgrounds.
Although our personal histories are all different, something
that felt true at the heart of each of our stories was that the
dominance of whiteness and internalised racism shaped our
early experiences, whether we grew up in the UK or not.
Interestingly, many of us shared the insights over 'third' spaces
(places we've lived outside of the UK or our 'home' countries),
where we gained more confidence and, critically, language to
negotiate our own specific diasporic identities.

Some ESEA people in diaspora assume that returning to an
ESEA majority space will provide immediate acceptance, comfort or reassurance. While there may be elements of truth to
this, it's not a perfect formula to rely on. Language, accent and

cultural nuances can become markers of difference rather than inclusion, reinforcing an unsettling reality: even within those 'home' countries, identity is categorised and scrutinised. The stark contrast between solidarity among ESEA people in the UK and the divisions observed in East and Southeast Asia points to a broader issue – identity is not only shaped by race but also by nationality, history and regional politics. The realisation that one could be an 'Other' even in spaces that seem familiar was a wake-up call, forcing a re-evaluation of what belonging truly means – if, indeed, it actually exists.

It's also important to consider the freedoms that different geographies can afford to those marginalised for their gender identity or sexuality. Most countries in East and Southeast Asia either directly criminalise non-heterosexual relationships, such as Myanmar, Indonesia and Malaysia (Human Dignity Trust, 2024), or do not have legal frameworks in place to protect and recognise the rights of LGBTQ+ people. The UK has provided many migrants with the freedom to express their sexuality and gender identity, but the lingering fear of being 'outed' may not fade easily after growing up in countries where expressing themselves fully is at best socially unacceptable, or at worst, dangerous[3].

Taken together, these perspectives reveal the intricacies of navigating multiple, shifting identities across different cultural landscapes. Whether through the comfort of community, the dissonance of feeling foreign in places that should feel

---

[3]We do also recognise that LGBTQ+ people, in particular trans and gender non-conforming people, are still vulnerable in the UK, and face the continual erosion of their gender-based rights to care and safety.

like home, or the challenges of expressing identity such as gender in different environments, the stories from our discussion underscored a fundamental truth: identity is never static. Instead, it is shaped by movement, experience and the constant negotiation between external perceptions and internal realities (Sen, 2006). Whether we feel it consciously or not, shared experiences ground us, and they help us feel at ease with the parts of us that can sometimes be sources of difference and Otherness.

LANGUAGE, CITIZENSHIP AND THE POLITICS OF BELONGING

For those navigating life between multiple cultures, citizenship and language are not just bureaucratic or linguistic concerns – they shape the very foundations of material living conditions, identity and a sense of belonging. In our conversation, co-author David articulated the weight of immigration status, describing the long process of securing indefinite leave to remain in the UK. The constant questioning of his right to stay had been an undercurrent of stress, influencing his decisions as a worker.

Language, too, functions as an unspoken passport, often determining social, professional and institutional inclusion. The markers of privilege are fluid; what signifies status in one country may be irrelevant (or even a disadvantage) in another. The way one speaks, presents oneself or even exhibits unconscious behaviours can be read differently depending on culturally-specific context, reminding us that identity is not just personally but also socially constructed and interpreted.

In the UK, accents play a particularly charged role in racialised belonging. For the children of immigrants, regional British accents can be a form of self-protection, with the knowledge that a British accent commands a certain level of respect sometimes denied to their parents. Their parents may speak multiple languages, but are often infantilised due to limited English proficiency or even merely due to racist *perceptions* of limited proficiency. Elders in particular are overlooked socially, leading to systemic neglect such as the underfunding of culturally specific community services or reduced access to social care (Age UK, 2020). Language barriers are not just about communication – they intersect with class, race and migration history to define who is heard and who is ignored.

Linguistic disconnect from a parent's mother tongue can create an unintentional distance from ESEA community spaces. We discussed the sense of exclusion for children who do not speak the language of their parents, and the empowerment of reclaiming in adulthood what was lost or denied in childhood: learning language, cooking traditional dishes, redefining traditional practices like ancestor veneration and reconnecting with family history on one's own terms. This part of our discussion is a testament to the idea that cultural identity is not fixed but can be actively shaped and recovered. Ever heard an inside joke using ESEA phrases or language but didn't know enough to follow along? While having a shared language ability can foster greater connection among diasporic ESEA people, in our experience, the opposite effect can happen when language ability hasn't been developed enough, as is common among many ESEA people we know living in the UK, where English dominates. Again, that feeling of not quite belonging can carry into experiences felt in ESEA countries where an accent immediately places you as an outsider.

Moving between different cultural spaces and constantly readjusting for the expectations of those around us is known as code switching, a term coined by linguist Lucy Shepherd Freeland (Freeland, 1951). Social media algorithms might have shown you the 'bye-lingual' memes that float around periodically, calling out those cross-cultural folks who recognise that their second or even third languages aren't quite up to scratch. Increasing numbers of social media content dedicated to the peculiarities of language hybrids like Chinglish or Vietglish emphasise the evolution of cross-cultural languages, once a source of embarrassment, but now a vehicle for diasporic solidarity (Thai, 2007). For co-author Amy, this process of reconnection took a particularly emotional turn when watching *Revolution of Our Times*, a 2021 documentary directed by Kiwi Chow about the Hong Kong protests. Hearing Cantonese spoken poetically and expressively reframed her relationship with Cantonese, revealing it as a bridge to both personal and collective history.

These experiences illustrate how language and citizenship are not just about legal status or linguistic ability, but also power as it relates to identity and emotional connection. A passport may dictate where one is allowed to live, but language determines how one is perceived, how much space one is given, and whether one feels at home.

We're often confronted with the vastly differing experiences of non-citizen ESEA people in typically lower paid, less secure sectors like the nail industry or domestic work, but such experiences are usually not centred in ESEA community activism. Indeed, one of the many weak points of identity-centred activism is a tendency to miss the complexities that create one set of material circumstances for one group and a different set for another, and the common threads across different marginalised groups beyond the 'identity' being centred.

While some organisations, such as Kalayaan, have been championing the rights of domestic workers for many years, it remains hugely important that migrant voices from all backgrounds be given space in any future organising in the movement for ESEA liberation from systems of oppression. Many second generation ESEA people in this movement are the children of working class immigrants, so ignoring the experiences of first generation workers in the fight for ESEA rights would be tantamount to the erasure of our very roots.

A PURPOSE BEYOND A LABEL

Sociologist Stuart Hall believed that 'cultural identity ... is a matter of "becoming" as well as "being". Far from being eternally fixed in some essentialized [sic] past, [it is] subject to the continuous play of history, culture and power' (Hall, 1987). Put simply, cultural identity, and by extension, ethnic identity, are not only constantly shifting, but are also an important part of shaping society and our role within it.

Ethnic identity, for racialised people and especially those living in majority white societies, is *always* political. Whether we exercise our ethnic identities consciously or not, as ESEA people, we are part of a critical cultural moment that is seeing a huge shift in the rejection of static, essentialist notions of race and behaviour. Instead, this movement is reshaping and course-correcting the meanings typically associated with ESEA communities – or, creating new ones entirely. While we all agreed that nurturing ESEA spaces provided us with the means to promote community healing and take meaningful collaborative actions, we also recognised that our lack of access to those spaces in the past was an important driver in our personal political pursuits.

Dr. Ruha Benjamin, Associate Professor of African American Studies at Princeton University, refers to this phenomenon – of connecting, sowing seeds and embedding daily practices into long term, sustainable systems – as 'viral justice' (Benjamin, 2022). She supports the position of Chinese American activist Grace Lee Boggs that 'the real engine of change is never "critical mass"; dramatic and systemic change always begins with "critical connections"' (Boggs & Kurashige, 2012). She also argues that every person, based on their skills, connections and passions, has a 'plot' of land in which they can sow the seeds of social change-making. Whether that plot is in the classroom, the boardroom, the kitchen or the garden, is down to each individual to figure out. It may seem simple, but revolutions are not won through everyone having the same skills.

ESEA people – or any people, for that matter – do not need to have identical experiences in order to create change, but creating connections through shared identities and interests allows us the freedom to identify and nurture our 'plot'. It's also important to recognise the ways in which some racialised people have felt excluded from white majority political organising spaces specifically due to a lack of consideration on how race and ethnicity impact material experiences. In the years since our group formed in 2020, the sheer volume of ESEA creative output we have seen, and the development of special interest groups engaged in political mobilisation, have created a rich tapestry of collective and individual experiences to inform and enrich not just our own understanding of our communities, but that of others, too. This makes for a useful foundation upon which to construct new forms of intergenerational, cross-cultural exchange that will be critical in any revolutionary movement.

Identity-based discussions between us, the co-authors, provoked an interrogation on why and how we go beyond surface level heritage exploration. For those of us who care passionately about interconnected global struggles and toppling oppressive apparatuses, we know that we must go beyond wanting ESEA people to thrive within broken systems. Instead, we must be prepared to commit ourselves to the creation of new systems, in the knowledge that this may be a multi-generational effort for years to come. Something we've learned in practice throughout the years is that widespread systemic change doesn't come with gradual behavioural social changes or even simply awareness-raising at institutional level. To give an example, our early days were concerned deeply with pandemic-related anti-ESEA violence, and the social marginalisation and exclusion experienced by ESEA people more generally.

We hoped that increased awareness-raising – which we achieved with nationwide engagement in the tens of thousands – would lead our institutions to implement radical policy change that would materially transform the lives of racialised people at the sharp end of the COVID-19 crisis. It didn't, unsurprisingly. We met with various departments and institutions that are meant to protect and promote the safeguarding of UK residents and citizens. Instead, we learned more of the ways in which our systems were imposing harm on the very communities we were supposed to be moving in solidarity with.

We also demanded more ESEA stories, more richness and diversity reflected in the cultural outputs and media we consume every day, and more diversity of experience among the professionals who are supposed to represent us democratically or care for our needs in schools, healthcare spaces and the workplace. If we had all of those things, would it impact the ethnicity pay gap, eliminate violence against women and LGBTQ+ people, or end racist immigration raids?

To be absolutely clear, we don't and should not overlook the value of significant changes in social and cultural attitudes when it comes to everyday experiences. The fact that more people these days do not condone racist slurs or preconceptions about ESEA people in workplaces, schools and social settings is undeniably a good thing. And, as we've discussed earlier in this chapter, having access to identity-focused communities and services, as well as seeing things you relate to reflected in the world around you, is an important part of feeling comfortable in your own skin *and* feeling ready to engage with political pursuits.

But how do we leverage those interpretations of heritage, of identity, for something more? How do we use them to practise and build alternative ways of doing, in preparation for the world we want?

For Yujie Zhu, a researcher whose work focuses on the intersection between heritage and tourism, religion, politics and memory, the ways in which heritage is interpreted in the public sphere, though complex and varied, can be viewed through a 'ladder' framework with five, non-linear rungs:

consumption and entertainment

knowledge and truth-telling

learning and understanding

imagination and immersion

reparation and reconciliation.

When we talk about 'heritage' in this context, it is easy to imagine a site or a physical place such as a museum or a site of historical importance, but it is just as necessary to consider the site of heritage interpretation among communities of people, whether in person or in digital spaces, as a matter of collective identity, whether imagined or otherwise. Zhu argues that heritage interpretation can be used as a form of public education, but caveats that this 'requires the needs of marginalized groups and communities to be recognized and prioritized [sic].' He also warns that our current systems of heritage interpretation, especially those which are institutionalised via bodies like the United Nations or through universities, rely strongly on 'professional, institutionalized [sic] knowledge and expertise to fulfil upper- or middle-class demands while ignoring local community needs. The heritage purported by these groups often reflects elite social practices and experiences' (Zhu, 2023).

What happens when we charge each individual within their sphere of influence with a responsibility to nurture and grow their own interpretation of heritage, and use it to build collective endeavours with others? To move past ESEA heritage interpretation as a simple means of consumption and any individual sense of belonging – recognising that it is much easier for some people to feel 'belonging' than others – we propose using identity exploration as a springboard to imagine new futures, working cross-culturally, intergenerationally and across class and border divides to demand and build those new futures. There is no point in imagining a world in which ESEA people can live free from racism if we do not also imagine a world in which all other communities are free from the harms of white supremacy complexes, border violence, economic instability, ableism and transphobia.

For this reason, it does not serve us to ask for 'equality' within the bounds of our current systems, as opposed to toppling those systems and creating new ones. While identity is a personal journey, we have seen how it is deeply intertwined with systems of power, and how it can be interpreted in the public sphere. If identity categories – both one's own experience of the world or personal characteristics, but also the perceptions imposed by others and shaped by historical systems – are intrinsically linked to oppression (Crenshaw, 1991), it follows that identity should also be a site of resistance. Identity is politicised, rather than essential. However, an over-focus on combating single political issues and surface level consideration of identity-based marginalisation, with no consideration of the systemic drivers of those issues and how they connect to other struggles globally, is sure to fail at really achieving any change for *all* marginalised people.

Intersectionality was first introduced as a framework for understanding – it was never the end goal. How do our identities inform how we act to neutralise the harms of systems like patriarchy, environmental colonialism and racial capitalism? How can we work with each other, using tools like intersectionality, to better understand our needs from a new society?

An individual cannot, of course, sustainably fight every battle on multiple fronts. To stay the course, organisers must know exactly what it is that they want to achieve. We must be able to question all of our organising methods, and be ready to adapt and change if the answer to the question 'Does this allow us in some way to achieve our goal?' is 'No'. We have mentioned many times that we don't have all the answers. Indeed, our own small organisation has seen its share of idea shifts, our vision has changed and grown and we have sometimes found ourselves trapped by pitfalls and in need of the very course-correction that we have mentioned above. In striving for perfection, we limit ourselves, but in allowing ourselves flexibility and adaptability, we strive for something better.

3

MIGRATION: HOW WE CAME TO BE

ROOTS AND ROUTES

This chapter makes some attempt to paint the broad lines of East and Southeast Asian migration to the UK. It will also look at the living, working and study conditions of some ESEA groups, and question the role of the state in shaping migrant experiences. We do not attempt to provide an exhaustive overview of every ESEA group in the UK, nor do we dismiss the complexity of ESEA migration. Instead, we encourage readers to reflect on the experiences detailed below, and consider how they apply in wider society or in their own spheres.

While there are recordings of early visitors from East Asia to Britain's shores, the first communities of ESEA people settled as a result of increasing global trade and colonisation. Between the 17th and 19th century, when the British Empire was at its height, Britain saw the arrival of lascars (sailors and militiamen from the Indian subcontinent and Southeast Asia, as well as East Asian Chinese and Japanese sailors). Many of these arrivals were Chinese sailors and naval cooks employed by the East India Company and the Blue Funnel Line, who often hailed from Kwangtung province. However, sailors from Chekiang, Malaya, Singapore and Fukien have also been recorded (The Old Bailey Proceedings Online Project, 2003). Communities were established in London and Liverpool, with migrants branching out into other trades such as laundry shops and restaurants.

In the 20th century, with post-war migration after World War II and the breakdown of the British Empire, the number of ESEA people coming to Britain for work and study increased significantly. The country also saw a growth in the mixed ESEA population. Migration to the UK has taken many forms – economic, study, asylum, marriage migration, to name a few – via both regular and irregular routes. Since we cannot dedicate sufficient space to each ESEA demographic, we have focused on a few case studies impacted by global politics, and the question of how we define and understand the 'we' of ESEA communities, as well as the socio-economic and political shifts that dictate their existence.

ESEA PEOPLE IN THE UK TODAY: ON PAPER

There has been something of a lack of consensus on how ESEA people refer to themselves (see Chapter 1: Who gets to be Asian?). This inconsistency has been compounded at a social and institutional level with regards to data collection and how visible our communities are. Whether or not that visibility in the eyes of the state is actually a good thing, we will discuss in due course.

In UK government censuses, since 2011, all ESEA groups other than 'Chinese' are recorded under the 'Other Asian' category (Cabinet Office, 2021). This blurring of lines is quite indicative of the treatment of ESEA people in society more generally. But perhaps we should be asking ourselves if data presented in this way can ever really capture the nuance of different ESEA backgrounds, which are extremely complex and varied. Ethnicity is broadly understood as a set of shared cultural characteristics, but even this is socially constructed and changes in different contexts. For example, 'Indian' is an ethnicity category in the UK, but within India itself there are thousands of ethnic groups – so 'Indian' as an ethnicity category in India is somewhat redundant.

The nature of ethnicity itself is complicated, which blurs our understanding when it comes to multinational, multiethnic population sizes. For example, in the UK censuses, many Singaporean people may select Chinese, Indian, or Other Asian as their ethnicity, meaning that birthplace is the only way of estimating their population size. This, by itself, is not an accurate representation, since many of those who identify as Singaporean may have been born elsewhere, and this number would also include non-Singaporeans born there.

One of the most important factors we need to consider when looking at ethnicity survey data, especially at national level, is the concept of respondent burden, or how hard, time-consuming, or emotionally stressful a person finds taking part in a survey. Things that can increase this burden include how long the survey is, how mentally challenging the questions are, how much effort is needed to answer, how often they are asked to participate, and whether the questions feel too personal or intrusive (Lavrakas, 2008).

Essentially, defining your ethnic, national and cultural identity for the purpose of fitting into a tick box is stressful, and can lead to non-response, attrition or inaccurate data.

In preparation for the 2021 Census in England and Wales, a survey on the previous Census found that those struggling with the ethnicity question included non-UK born individuals, non-English speakers and those unfamiliar with ethnic categorisation. Some gave different answers in the survey than in the Census, especially between 'Asian' and 'Mixed' identities (ONS Census Transformation Programme, 2016). This highlights confusion among mixed-background individuals, as 'Asian' in the UK typically refers to South Asians, while 'Mixed' often implies Black/white heritage.

So why are we doing it?

Furthermore, we must be aware of how data is stored and used with regard to historically excluded groups, and whom it serves. Even if data is accurate, how it is presented has the potential to prop up particular narrative biases. After the 2021 Census in England and Wales, some politicians and newspapers seized on the minor decrease in the number of white ethnicities reported (86% to 82%) (Office for National Statistics (ONS), 2022), and the decline in those identifying as Christian, to bolster the claims of the Great Replacement Theory (Hamdache, 2022)[1]. Alongside a general rise in Islamophobia, it is easy to see how this data misuse is worrying at best and dangerous at worst.

---

[1] The Great Replacement Theory is a conspiracy theory that claims white populations are being deliberately replaced by immigrants through migration, higher birth rates and cultural changes.

In addition, our lack of disaggregated data at national level makes it difficult to understand the specific issues faced by particular communities, making it easier for generalisations that prop up the model minority myth to emerge. For example, claims that Asian migrants are all financially stable and benefit from higher education are without evidence once we start looking at disaggregated data. If we look at South Asian ethnic groups' median hourly pay from ONS figures from 2022 (ONS, 2023), we can see that the difference between Indian and Pakistani or Bangladeshi ethnic groups is significant. Indeed, between Chinese and 'Other Asian' groups, we also see a difference, and these averages may be skewed by extreme numbers at both ends. We must assume that the 'Other Asian' category includes, for example, working class Filipino and Vietnamese migrants, who are more likely to have lower paid jobs such as domestic workers and nail technicians.

There are also risks in leaning on data to perpetuate stereotypes that completely ignore the history of colonial legacy and present day exploitation in parts of East and Southeast Asia, which drive migration to the Global North. ESEA people and migrants in general are also more likely to support family members 'back home', allowing them less disposable income and social mobility (Banfield-Nwachi, 2022). 'Official' statistics tell a small part of the whole story, and it's important we view them critically, alongside important research from non-governmental organisations, and the testimony of people's own experiences.

We must continue to question and push back against the ways in which the state and other institutions use data, for example in the case of the controversial Police, Crime, Sentencing and Courts Act of 2022. Numerous groups, from abolitionist collectives to human rights organisations, warned of the dangers of the Act at inception. Advocacy group Liberty claimed it had potential to 'restrict protest, criminalise trespass and erode limits on police powers to collect big data, profile people in the pre-criminal space, and subject people to suspicionless stop and search' (Liberty, 2021). Other efforts to restrict and combat the surveillance of the state through biased datasets include those of Erase the Database Collective, which encourages Black Londoners to check whether their data has been included in the Gangs Matrix – a framework for categorisation formerly used by the Metropolitan police following the 2011 London riots. It is based on vague criteria like postcodes, music tastes and friendship connections. Numerous evidence compiled by projects like Erase the Database suggests that the Matrix led to over policing of Black communities in London, increased evictions, school exclusions, restrictions on driving licencese and deportations. It was found to be discriminatory following a legal challenge (Amnesty International, 2018).

While data is important for making sound policy decisions, it should not be considered simply as a 'receipt' that must be produced to prove experiences of discrimination or racism among marginalised people. Nor is it a tool for institutions like the state to scapegoat, exploit or isolate vulnerable migrant groups. Research that we undertake to better understand our communities should always have community care as its focus. Sadly, this is not often the case when it comes to government data.

AN ETHNIC IDENTIFIER IN THE MAKING

As the ESEA identifier becomes more popular, we are starting to see research and documentation on its origins and evolution.

According to Jonathan Gray, a researcher in digital data, methods and infrastructure in society, the ESEA acronym emerged as early as 2018 as a 'bottom-up collective ethnic identifier' (Public Data Lab, 2021) for East and Southeast Asians. According to Gray and other researchers at the Public Data Lab, there are three key uses for this acronym in the 21st century:

> In solidaristic response to the intensification of anti-Asian racism and violence spurred by the COVID-19 pandemic

> To campaign for the institutional inclusion of ESEA communities

> As a form of political community-building

If we are witnessing an identifier in the making, then we also have a chance to actively shape the way we evolve this term. It's important that we recognise language's ability to shift in response to the changing needs and dynamics of the current social and political context. In the same way that the words we used 50 years ago may no longer be appropriate for our discussions on race, class and ethnicity today, it may equally be the case that new thinking and developments in the field will soon necessitate a change in terminology.

For now, though, mapping the steps of those who choose to rally to the ESEA identifier, as well as those who don't, provides us with a new understanding of what it means to choose, live, perform and sometimes reject a political identity. This political identity connects people across commonly shared backgrounds from a unique cross section of Asian diasporic and transnational identities in the UK. Although, as a singular label, it retains certain limitations – as any term would – and cannot fully encapsulate all the demographics of the region, it allows us to refer to people whose origins share cultural and ethnic markers without making reference to skin colour or religion.

EAST OF WHERE?

'British' is a common identity prefix we often see in discussions of ESEA people in the UK, such as 'British Vietnamese' or even 'British Born Chinese'. Despite the fact that one does not need to have been born in the UK to be considered British *and* Chinese, the use of the qualifying adjective is interesting. Professor and researcher Gregory Lee makes an excellent point about the Othering nature of the term 'British Chinese'. 'British' is a qualifying adjective for a Chinese person, whereas with terms like 'Asian American' (the battle for which was hard fought by Asian American activists and academics), 'Asian' is a qualifying adjective for an American person. Asian Americans are American first, not the other way round[2] (Lee, 2021).

According to Lee, the eurocentrism of the globe and the domination of 'Western' modes of thinking in research, policy and communications will limit our ability to truly 'decolonise' education with regards to East Asian studies, because the modes themselves are rooted in colonialism. The use of 'our' here is appropriate, as we are a group of writers all resident and educated in the UK or systems of British colonial influence.

---

[2]At least, this was the idea behind the terminology. The reality may be quite different.

Lee argues that the term 'East Asia' and the field of East Asian studies 'owe their existence to this one-way Euro-American vision of the world. Thus, when the terms Oriental and Far Eastern were deemed to reek of the imperialist era, they were replaced by the more modern-sounding but equally politically problematic neo-colonialist designation of East Asia.' Lee also suggests that people nowadays feel more comfortable with 'East Asian' as a more inclusive term to avoid sinocentrism, admitting that 'ultimately, there are few palatable ways out of the dilemma "bequeathed" to us by British colonialism and its afterlives' (Lee, 2021).

With the understanding that we have limited 'ways out' of the colonial imposition, we should also acknowledge that using 'ESEA' as a more inclusive term doesn't automatically eradicate exclusion and discrimination among our communities. However, it's certainly a welcome alternative to the now outdated acronym 'BAME'. Many ESEA people, particularly those who are especially marginalised along different intersectional axes of ethnicity, migration status, economic status, disability status or gender identity, are still excluded from mainstream conversations and spaces supposedly inclusive of all ESEA folks.

It's also naive to assume that we'll all get along all the time or agree on every issue, particularly when it comes to politics, which can be a dividing line that impacts individuals' engagement with the dynamics of race and class. People may feel a stronger affiliation with groups or communities that share other identities or beliefs, such as LGBTQ+ identity or faith. Some people, especially those from older generations, may not even remotely identify with the term 'ESEA'. Terms such as 'oriental' and 'yellow' – which have their origins more strongly rooted in white European colonialism and racism, and therefore are widely rejected – are still in use by some, and even reclaimed as a political identity, in resistance to the palatability required for inclusion within harmful, carceral institutions and practices (Nuallak & *Remember & Resist*, 2024).

Whatever the language of self-identification, we encourage our readers to explore new possibilities and critically evaluate their choices with intention, especially when it comes to their own identity expression and how they move in solidarity with others.

MIGRATION PATTERNS

Diasporics are moving targets. What they are today, they will not be tomorrow. (Watson, 2004)

When anthropologist and Fairbank Professor of Chinese Society, James Watson, wrote these words, he was reflecting on 35 years of research on a 'tightly bound kinship group' of migrants from San Tin in Hong Kong's New Territories. The claim rings true for most first generation migrants in Western diaspora. These people have always been at the constant mercy of policy and legislative changes to immigration, media attitudes towards migrants, social perceptions of diaspora communities and interpersonal interactions – which affect where people move, and how they live and work. These variables also impact their sense of self and attachment to their ethnic, cultural and national identities.

As we will see in this chapter, ESEA presence in the UK has been impacted by immigration policy since World War II, for both the migrants who live and work within the bounds of the establishment, and for those pushed to the outside, many of whom represent some of the most vulnerable people in our society.

The British Nationality Act of 1948 is now widely understood to be partly responsible for the growth in diverse demographics in the UK up until the early 1980s. It was designed to fill gaps in the labour market after World War II, and granted the status of 'Citizen of the United Kingdom and Colonies' (CUKC) to those who had been born in British colonies. The British Empire, at its height, was considered to be the largest empire in history, covering almost a quarter of the Earth's land area. Britain's colonies in East and Southeast Asia included: Burma; Malaya, Sabah and Sarawak (all now modern-day Malaysia); Singapore; Brunei; and Hong Kong.

While most of us largely understand the Act to have led to the establishment of the Windrush generation (CUKCs from the Caribbean), it also led to the arrival of more CUKCs from Africa and Asia, particularly from Hong Kong. Taking advantage of economic prosperity and the rise in dining out, many Hong Kong Chinese families flexed their culinary muscles by opening takeaways and restaurants, introducing wok hei to the British masses (not without a side of pies and chips in order to appeal to local tastes). To avoid competition with similar establishments, restaurants and takeaways were often located far from each other – contributing to the isolation felt by the families running the businesses.

Sometimes, however, the isolation of ESEA communities was a result of deliberate government intervention. The Government Dispersal Policy of 1978, which applied to the so-called Boat

People[3], stated that refugees should not be allowed to cluster in large numbers in particular areas. Dispersal was to be a central component of the resettlement process. This meant that some 22,000 refugees – their lives upended and suffering the trauma of their escape from conflict – were separated and housed in smaller numbers across the UK (Taylor et al., 2021).

As early as spring 1980, the Joint Committee for Refugees from Vietnam (JCRV) advisors began flagging up the dangers of isolation, and how the government's policy of avoiding 'ghettos' had made it almost impossible to provide effective language tuition for refugees. With only two or three families in an area, it was neither financially nor practically viable to deliver dedicated English classes to such few students. A survey by Ockenden of 1,530 refugees living in its resettlement zones found nearly half living three or more miles from their nearest English tuition class, with nine people living over fifty miles away (Taylor et al., 2021). Housing shortages and rising unemployment led many people to move to larger cities, principally London.

Despite the veneer of compassion offered by the Thatcher government to refugees seeking safe haven from the ravages of conflict, it failed to provide the charities and community centres with much needed logistical support to help people carrying the trauma of war and displacement to settle with ease (Kushner, 2006). Furthermore, the refugee population was fractured along lines of ethnicity, politics, faith and

---

[3]Refugees from Southeast Asia, namely from Laos, Cambodia and Vietnam. These included large numbers of Viet Hoa, ethnically Chinese people who lived in Vietnam and who suffered political persecution there.

class, meaning that even close resettlement did not result in automatic community cohesion (Taylor et al., 2021). In an interview for the 2021 digital storytelling exhibition *Land, Sea and Stars*, Huong Black, daughter of Viet Hoa refugees from Vietnam, explained that these fractures impacted her parents' ability to connect with many other Vietnamese people at what was undoubtedly a very difficult time in their lives. In turn, this has a knock-on effect on the children of the impacted people, as one example of what we now understand as intergenerational trauma (Warner, 2021b).

Intergenerational trauma, first recognised in the children of Holocaust survivors, is when the consequences of a traumatising event are passed on via the shared family environment to the children of the direct victim (Yehuda & Lehrner, 2018). Similarly, transgenerational trauma is the collective traumatising event experienced by groups of people sharing a common identity: the Holocaust, the transatlantic trafficking of enslaved people, the Vietnam War, and South African apartheid are all examples of this. In the case of migration from Vietnam, prioritising integration had a fracturing impact on relationships between older first generations and younger first generation or second generation migrants. While the newer generations lost their mother tongue, they also lost an ability to communicate meaningfully with their parents and grandparents on the grief and trauma of displacement (Tingvold et al., 2012).

While many second and third generation children and grandchildren of former refugees from Vietnam, Cambodia and Laos are taking steps to identify and reckon with their inter- and transgenerational trauma, there remains a dearth of outreach and services for older generations who do not find it easy to approach topics of mental health, even though their need may be even greater. Jack Shieh, former director of Vietnamese Mental Health Services (VMHS), an organisation that provided services for Vietnamese people with mental health needs, says that there is a huge need for more Vietnamese people from the younger generations to train as professional psychologists (Warner, 2021b). As it stands, much of the counselling done among Vietnamese communities is informal, providing practical rather than clinical support. Sadly, VMHS, the only organisation of its kind, closed in September 2023 due to withdrawal of funding. Potential and previous users of the service will be re-routed into the wider NHS system, which lacks the culturally-aware care previously offered by VHMS.

Like the Chinese restaurant and takeaway workers who ushered in a new evolution of British Chinese cuisine by adapting their food menus to suit local palates, so too did refugees from former Indochina adapt to their new surroundings in order to survive. Many Viet Hoa people found work in the Cantonese speaking kitchens of Chinese establishments while other Vietnamese found work in garment factories, or sewed clothes by hand in their homes. However, as the clothing industry in the UK began to decline – with companies outsourcing labour abroad in the 1990s – more and more Vietnamese turned to new forms of entrepreneurship, most popularly in restaurants and nail salons (Wilkins, 2021).

More recent movements of Vietnamese people into Britain fall into the categories of asylum seekers, irregular migrants, overseas students and – to use the Home Office's own parlance – 'highly skilled' professionals. Some Vietnamese community organisations have estimated that there may be up to 20,000 irregular migrants living in the UK (Sims, 2006). Vietnam is the second highest source country for modern slavery victims in the UK (University of Nottingham Rights Lab, 2021). It has been estimated by the United Nations Office on Drugs and Crime that 30 Vietnamese children arrive in the UK every month (Kelly & McNamara, 2015), forced into a life of slavery and exploitation, with many of them ending up on cannabis farms. Mainstream media has done a good job of redirecting the narrative on responsible 'Vietnamese gangs' to one that bolsters negative social perceptions on the criminality of migrants, irregular or otherwise.

As is often the case, such narratives stem in part from institutional biases, and we can see these ideas being perpetuated explicitly by the police. The West Yorkshire Police Force drugs coordinator, Bryan Dent, said in a 2006 statement following the arrest of seven Vietnamese people from cannabis dens, 'the Metropolitan Police have had a *Vietnamese problem* for some time and maybe *their people* think they can go about their business in relative anonymity in *our city centres*' (Casci, 2006, italics added). In the London Borough of Barnet's official Metropolitan Police guidance for landlords, links between the Vietnamese and cannabis cultivation is made suspiciously plain as it states, 'Almost invariably residents of these [premises used as cannabis factories] will be of Vietnamese origin,' and, 'estate agents and landlords should beware of lone females, possibly Vietnamese, trying to rent property' (Sims, 2006). While data on Vietnamese people being trafficked into the drugs trade in the UK ranges over a 20 year period, the organisation Anti-Slavery International maintains the issue is still prevalent.

In a 2006 Runnymede report, Jessica Mai Sims, a policy and research professional, claimed that the police made no attempt to distinguish between innocent or guilty, or between settled Vietnamese people and those newly arrived. Sims claimed, 'In advising landlords to beware of Vietnamese individuals trying to rent property, the police fall short of meeting their obligation to have due regard to the need to, "eliminate racial discrimination" or "promote equality of opportunity and good relations," as required in the Race Relations (Amendment) Act 2000' (Sims, 2006). Under the guise of acting in the public interest, the police have been shown to fan the flames of systemic Othering and racism, raising the question: who is being protected? And from what?

HISTORY BOOKS, BUT FOR WHOSE HISTORY?

Britain's history is littered with examples of those who have been abandoned, forgotten and mistreated. Our school history curricula and what is remembered in the public consciousness have always been a choice about which stories are included. This raises questions about the role of the chooser: if those responsible for documenting, elevating and teaching stories are only interested in eurocentric narratives, where does that leave racialised communities? In our research, we noticed a general tendency for areas of study – not always, but often – to be conducted by people with personal connections to the groups or topics studied (in our case, ESEA writers, researchers and academics). Some find themselves caught in a bind that many writers and researchers of colour experience: they want to be the ones to tell our stories, but they don't want to be exclusively consigned to the labour of advocacy.

In schools, British students are taught in great detail about World Wars I and II and the way they shaped modern geopolitics, but there are so many details that escape our pedagogy – such as the internment of Japanese people on the

Isle of Man or the reliance on troops from the British colonies. After the outbreak of World War II, British residents who were nationals of Axis Power countries, namely Germany, Italy and Japan, were interned in camps on the Isle of Man under suspicion of affiliation with enemy states. Many of these people had been living in Britain since childhood and, in the case of many German and Austrian prisoners, were Jews who had fled antisemitic persecution and were highly unlikely to pose any threat. However, very little is known about the Japanese prisoners (Pistol, 2017). How were they treated by the guards and their European fellow prisoners? When were they released and what happened to them afterwards?

These gaps in research and documentation may even mean that this knowledge is lost to us. During research for the documentary film *The Six* (2021), filmmakers Steven Schwankert and Arthur Jones felt similar levels of frustration in their quest to uncover the stories of the six Chinese survivors of the Titanic, who disappeared under a shroud of media gossip mongering and discrimination. Much is known about the survivors of our most famous shipwreck, with the exception of these six men, who seem to have been written out of history. Facing deportation upon arrival, in the grip of Chinese Exclusion law, they appear to have simply vanished.

No lower education in British schools is complete without an appreciation of the sacrifice made by British soldiers in the World War I, the plight of evacuee children, or a visit to the Imperial War Museum for a day spent in imitation trenches. These histories are undoubtedly important, yet they often overlook the contributions of the Chinese Labour Corps (CLC) – a significant force of 140,000 Chinese labourers who left their remote villages in rural China to provide essential support services behind the frontlines for the Allies (Xu, 2011).[4] Their role was to provide the support work and manual labour that would free up British and French troops for fighting on the front lines.

Recognition of their service has been almost non-existent in Britain, especially as these men were refused the right to settle and therefore do not have descendants here. The British government did not actively promote or preserve any information about the CLC, and out of the 40,000 WWI memorials in the UK, only one was created for these men – at China Exchange in London, which has now unfortunately closed. Efforts have been made in recent years to uncover and bring attention to this erased history, including the 'Ensuring We Remember' campaign founded by the Chair of the Chinese in Britain Forum, Steve Lau, and a 2018 play by Daniel York Loh, entitled *The Forgotten*.

It is also crucial to understand this story within the broader context of Britain's complicity in the violence of empires that supplied armies drawn from across the globe. The tendency to present Britain as one of the 'good guys', without meaningful critique or context, is a common feature of British education. While we rightly acknowledge the struggle and sacrifice of soldiers from Britain, the Caribbean, India, Africa and China, among others, we must also recognise that imperialism and the harms of a military-industrial complex should not be justified in their names.

After World War I, many discharged soldiers from colonised countries and Chinese labourers settled in Britain's port towns, where they were met with hostility and racism by local

---

[4]It should be noted here that the troops who served from the Global Majority in World War II were referred to as the British Colonial Auxiliary Forces – 'auxiliary' implying support and subservience rather than agency or empowerment.

populations as they were seen to be undercutting local wages. Tensions rose, particularly in Wales, resulting ultimately in the Race Riots of 1919. Racist and xenophobic sentiments had been in place prior to the war, with the 1911 Cardiff Seamen's Strike focused on Chinese sailors, resulting in all of Cardiff's laundries (run by Chinese migrants) being vandalised (Evans, 1980). In the 1919 riots, anyone perceived as not British, that is to say, not white, was a target. The front page of the *Cambria Daily Leader* from 14 June 1919 gives this description of some of the victims:

> The threatening crowd burst in upon the shop of a Malay boarding-house keeper. Some of the occupants took flight, and by means of a rope thrown around the chimney they reached the roof. When dusky forms were seen on the skyline, jeers went up from the crowd, and a shower of stones and missiles were thrown. (*Cambria Daily Leader*, 1919)

Reading this account, one is reminded of the current wave of anti-migrant sentiment present in the UK at the time of writing, with far-right attacks on hotels housing asylum seekers increasingly common under the Johnson, Truss and Sunak governments (Townsend, 2023).

What is also tragic is to see history repeat itself in the aftermath of World War II. In the years since 1945, Allied countries have all dedicated significant effort into remembrance of all those who served. However, the contributions of Britain's colonial troops have rarely been spoken about. Over two million Asians, Africans, Caribbeans and Arabs fought and died for the Allies in the war (Olusoga, 2016), but their legacy has been a quiet one. One particularly heinous act of injustice meted out by the British government after WWII concerns the fate of the

Chinese seamen residents of Liverpool, a story that would only really come to light decades later with the declassification of HO/213/926. The contents of this file were top secret, not to be acknowledged in public. The title? 'Compulsory repatriation of undesirable Chinese seamen'.

Some 20,000 Chinese men were in residence in Liverpool, having been employed by the Blue Funnel Company as sailors and cooks. During World War II, these men 'kept the British merchant navy afloat, and thus kept the people of Britain fuelled and fed while the Nazis attempted to choke off the country's supply lines. [They] were a vital part of the allied war effort' (Hancox, 2021). When the British government secretly and forcibly deported these 'undesirable', unwelcome men from Liverpool in 1945, they left behind English wives and mixed children, who had no clue what had happened to them. A whole generation of mixed Chinese Liverpudlian children grew up feeling robbed of a key part of their identity, bullied for being different, gaslit by subsequent governments and excluded from British Chinese communities.

Since migration routes focused mainly on economic activity, which favoured emigration of men rather than women (Benton, 2007), Chinese seamen in the UK often settled down with white British women. The community network organisation Dragons and Lions, founded by Yvonne Foley, who is herself a child of a deportee, points out the stark contrast with the privileges afforded to white residents of China during the same period. The organisation has spoken out against the treatment of these children and their mothers by the State, which denied and removed British citizenship (if the mothers were legally married to Chinese 'aliens'), as well as the rejection of the children by white society, Chinese communities and even their own families (Dragons and Lions, 2021).

In the case of the Liverpool seamen, we see again how policies are manipulated to suit racist, anti-migrant agendas.

The British government, set on deporting these men, was faced with the challenge that none of them were actually in residence 'illegally'. The only grounds for deportation would be if they were in breach of their landing conditions, which they weren't. The solution? Vary the landing conditions to require the men to leave shore two days before their actual departure date, thus causing a larger number of 'overstayers' and giving the Home Office legal grounds for arrest and deportation (Lee, 2021).

It is in spite of, rather than with the help of, British governments that the stories of these men and their families have come to light, although most of the fates of the deported seamen themselves may be lost forever. It is a testament to grassroots community organising that there is now a commemorative plaque at Liverpool's Pier Head, a BBC documentary, drama and poetry dedicated to this history. It is a campaign that continues, seeking a government apology for this heartbreaking betrayal.

While it's important for the government to be accountable for its actions and for the impacted community to seek redress, justice will not be a straightforward affair, and ESEA community organisers must play their part. We have welcomed initiatives in recent years to include the stories of the Liverpool deportations and the families left behind in ESEA community building and organising, such as the panel event hosted by the British Chinese Society in London in 2023, and in Manchester in 2024 respectively for ESEA Heritage Month.

PLACES: WHERE WE GATHER

As we explore in Chapter 5 on mixedness and multiculturalism, spaces for social interaction and exchange of goods, services and knowledge – whether physical, or digital – have a huge role to play in supporting the development and longevity of communities. Notable mini enclaves[5] of ESEA people in the UK include the Chinatowns of Liverpool, Birmingham, Manchester and London, the Edinburgh neighbourhood of Newington, which supports a strong ESEA student population, the London boroughs of Lewisham and Hackney for the Vietnamese and 'Little Manila' in London's Earls Court, to name just a few. As much as people in diaspora are at the mercy of institutions and policies, so too are the spaces in which they choose to gather subject to shifts in council funding, demographic politics and consumer behaviour. Places occupy moments of history in ever-changing capacities, and ESEA spaces are no exception.

British expansion, thanks to merchant shipping companies, meant that London's East End dockside neighbourhoods were a natural site for what is commonly understood now to be the first London Chinatown (distinct and separate from today's Chinatown, which is found in Soho). By 1890, there were two very small yet stable enclaves of Chinese people in the riverside district of East London: those from Shanghai around Pennyfields and those from Guangzhou (Canton) and Southern China on Limehouse Causeway (Benton & Gomez, 2007). We have already mentioned the hostility faced by post-World War I settlers in Britain's port towns, and the destruction of Chinese laundries and immigrant businesses. While we may be, sadly, unsurprised at the accounts of racism and violence towards racialised migrant groups,[6] it is important to contextualise

---

[5]Mini, because dispersal of communities in the UK means that we do not have true enclaves in the same way that cities like Paris or Los Angeles do.

[6]It should be noted that migrant groups such as Poles, Romanians and Roma have experienced significant hostility and Othering with varying intensity across Western Europe, particularly in the wake of Brexit.

these incidents within popular culture. Alongside politicians stoking fears of the threat of cheap migrant labour, some significant and enduring myths of Yellow Peril began to take root in the public's imagination through exaggerated reports of illicit drug trade, gambling, opium dens and the ensnarement of 'innocent white women' by 'Chinamen' in the sensationalist press (Stone, 2017). This was bolstered by works of fiction like Thomas Burke's *Limehouse Nights* (1916), and further fuelled by those wishing to capitalise on the dramatic potential of 'Chinatown'.

Chinatown became an ubiquitous cultural reference in novels, film, theatre and songs, from Charles Dickens' disparaging descriptions of Limehouse's squalid opium dens to Arthur Sarsfield Ward's (Sax Rohmer) creation of Dr Fu Manchu, the ever resilient and enduring embodiment of Western sinophobia, or as Rohmer himself put it, 'the Yellow Peril incarnate in one man' (Rohmer, 1913). During that same period, although perhaps not so well known as the likes of Burke and Rohmer, we are offered balanced, nuanced accounts of Chinese experiences in London through the work of writers like Lao She, whose 1929 novel *Mr Ma and Son* explores Western prejudice in conflict with Eastern stoicism.

Against this backdrop of social exclusion from British society, Limehouse's Chinese population was dispersed by a series of deportations and police raids aided by government policy such as the Defence of the Realm Act (1914), which criminalised opium use and gave the authorities increased powers to deport Chinese people and restrict their ability to work. There were also slum clearance schemes that sought to redevelop low-income areas, and by the 1930s, London's first Chinatown was no more (Witchard, 2017).

Today's Chinatown in London is nestled in the vibrant neighbourhood of Soho, away from the roaring traffic of Shaftesbury Avenue which attracted post-World War II Chinese migrants for its cheap rents and already thriving migrant population. By this point, British soldiers who had served in East and Southeast Asia now had familiarity with – and cravings for – eastern flavours, and the hospitality trade thrived. These days, Chinatown serves not only as a hub for people with origins in Mainland China and Hong Kong, but for other ESEA communities whose roots are from the likes of Taiwan, Malaysia, Indonesia, Singapore, Hong Kong, Korea and Japan.

It's also a popular tourist attraction, and its restaurants satisfy the tastebuds of non-ESEA visitors, while also providing a convenient hub of ESEA supermarkets, bakeries and supplies – still loved by ESEA communities today, though less critical than its earlier years, when it acted as a community lifeline for dispersed families. Other than a commercial centre, Chinatown has also become the site of political focus in the past couple of decades, with particular attention on rising costs and gentrification (Lucas, 2015). The pandemic saw calls online for people to support local restaurants and takeaways by ordering deliveries, but it was an especially difficult period for Chinatown businesses, as Soho is no longer a very residential area. Businesses had low outputs and high rents to pay.

Today, there are many questions about who speaks for Chinatown, and who Chinatown is for: the businesses it houses, or the consumers it caters to? Organisations like the now closed charity China Exchange, formerly on Chinatown's iconic Gerrard Street, as well as its successor, the community-based group Chinatown Collective, have aimed to provide a space for intentional community building, as well as a place to connect those who wish to learn about China, Chinese culture, and Chinatown specifically. Through a small team and a network of volunteers, the charity hosted festivals, exhibitions and heritage safeguarding projects such as historic walking tours.

More ESEA communities have gathered in other areas, notably in London and other large cities, which offer more jobs and reliable business for entrepreneurs. The southwest London suburb of New Malden was the site of Samsung's first European headquarters in the 1970s and the original South Korean embassy, which led to the emigration and settlement of a thriving Korean community that makes up around a third of New Malden's population. It is the cultural heartland of the Korean community of London, providing supermarkets, restaurants, karaoke bars and Korean language churches and schools. The large numbers of Koreans already settled by the 1980s – with much business and trade spoken and written in Korean – made it easier for newer arrivals, including a small community of North Korean refugees. However, similarly to the early Vietnamese populations of Britain, geopolitical factors mean there are clear divides between South and North Koreans, supported by urban myths and propaganda from back home (Fischer, 2015).

A SHARED TABLE

Food is widely understood to be of critical importance to community cohesion and relationship building for diasporic peoples the world over. For ESEA people in Britain, it can provide a connection to the homes left behind, employment and places to gather and socialise, with many people relying on their interactions at specialty supermarkets as their source of community knowledge exchange. Food offers the opportunity for identity exploration, intercultural and cross cultural exchange, and connection. Naturally, most businesses in Chinatowns and ESEA mini enclaves centre around food: supermarkets, restaurants, cafes and dessert shops.

One recent example of a vibrant food scene springing up in recent years is the south Edinburgh neighbourhood of Newington. Home to the Central Mosque and the main University of Edinburgh campus, it has a large Asian population. At the university, according to its 2022–23 intake, 54% of students were from overseas, with 20% of the total student population domiciled in China, Hong Kong, Malaysia, Singapore, Indonesia and Japan alone (The University of Edinburgh, 2023). Within the last decade, the number of ESEA businesses surrounding the university has skyrocketed, without a doubt to satisfy the needs of the international student population (and delight other, non-student residents).

Writer, PhD researcher and former international student advisor, Vy-liam Ng, spoke to us about how ill-equipped many universities are to support their ESEA student populations, thereby reinforcing the need for cultural spaces in the local area. In addition to institutional shortcomings, stereotypes and assumptions made by white students about Asian students, he believes that the differences in cultural practices involved in student bonding are simply not addressed. While most Asian students seem to prefer gathering over food, American and British students are deeply entrenched in a culture of drinking, making students who are Asian nationals more likely to socialise with other overseas students (V.-L. Ng, personal communication, 7 August 2023)[7].

---

[7]When we say 'British', 'American' and 'Asian' in this context, we are, of course, referring to nationalities.

However, it should be noted that, as with most other diasporic communities the world over, equating ESEA identities with more whimsical things like noodles or bubble tea – while it may provide light comic relief in community organising spaces – does not represent true identity exploration and risks flattening the richness of ESEA food in diaspora. In the United States, activists have been critical of so-called 'boba liberals'[8], who are often viewed as using surface-level tropes to assert their racial credentials within the neoliberal status quo, but then reject more progressive movements that address the issues facing Asian Americans, including the material living and working conditions of the most marginalised (Frias, 2021).

This is not to say that the history of food and its preparation is whimsical, of course – rather the opposite. There's also nothing inherently wrong with a popular culture that embraces connections over shared tastes and experiences, but let that not be the end of our conversations around identity. Rather, we should look at the benefits for community members – such as one journalist observed at the formation of a new ESEA space in Edinburgh's Newington. As well as helping stem student homesickness, these establishments allow people from different countries and regions to meet and explore identity through food, and they provide restaurant owners spaces to showcase their regional cuisines proudly and ground them in a local community (Kwek, 2024).

PRESENCE BEYOND THE SYMBOLIC

Some distinctive features of Chinatowns in the UK are their landmarks, such as the arches of Chinatowns in Manchester

---

[8]In the USA, bubble tea is commonly referred to as boba.

and London, or the pagoda of Birmingham's Holloway Circus. These landmarks act as beacons to announce ESEA presence, as if to say, 'We are here.' The symbolism of how landmarks are chosen and erected should not be ignored: the Manchester Chinatown arch, for example, was a gift from the City of Manchester Council to the Chinese community in 1986. Gestures like these, while they may only be symbolic, can be a significant step towards inclusion at public and institutional level. In an increasingly diverse Britain whose councils want to pay homage to its rich heritage through public installations such as the blue plaques scheme, perhaps there is further need for public recognition of our communities.

At the time of writing, we have just seen the lights of Oxford Street in London lit up for the first time ever for Ramadan and Mayor Sadiq Khan's Untold Stories initiative – a £1 million fund to create public artefacts to uncover the stories of historically marginalised communities – which promises to afford the capital's diverse population the visibility, recognition and public celebration it deserves (Greater London Authority, 2021). However, as with any act of representation or 'diversity' initiative, we should be wary of surface level inclusion. Symbolic gifts mean nothing if they are not supported with fair and equitable systems that include and support marginalised people in society. Even within government remits, a four or five year term that passes in the blink of an eye, policy reform is a band-aid on a bullet wound under an economic system that values profit over people.

Space itself is a commodity, and the existing spaces are hard won against the looming shadow of capitalism. Gregory Lee writes of his decades-long work in fighting for spaces for Chinatown communities to ground their heritage exploration and display their histories, expressing regret that so-called 'business communities' often reject local community initiatives that require funding, prioritising instead initiatives that encourage more 'authentic' overseas Chinese money and trade: 'The local in the UK context belongs to that blighted, ever-present, post-colonial "legacy" deriving from Britain's nineteenth-century imperialist adventures in what is now China and south-east Asia' (Lee, G., 2023).

It is high time that Britain's institutions squared up to their colonial legacy in tangible, meaningful ways to support impacted communities. It is also time for individuals and communities to reject superficial notions of 'authenticity' (often driven by business interests) and embrace new ways of being: ways that are imperfect, perhaps, or fragmented, hybrid, patchwork and evolutionary in nature. While they are unlikely to make meaningful material change on their own, some of these new public initiatives do offer valuable opportunities for individual community members to meet and organise together, building the foundations of care and support networks that are so desperately needed for sustainable organising.

WHERE ARE WE GOING?

In a country where political and legacy media narratives seem to be dominated by narratives around invasions, small boats, and suspicion of 'foreign born' people (Carlile & Harrison, 2022), to be an immigrant is to be at the mercy of borders and the often violent ways in which they are managed and policed. In 2022, the controversial Nationality and Borders Act made it possible for the British government to remove a British-born person's citizenship for 'the public good' without informing them (previous iterations of this power did not apply to born citizens and required notice to be served). Consider that the same government adopted the Police, Crime, Sentencing and Courts Act in 2022, which significantly restricted rights to protest and granted more powers to the police. This saw MPs and ministers equating demonstrations of Palestinian solidarity with terrorism and hate marches (Syal et al., 2023), and illustrates how much immigrants and the children of immigrants stand to lose for not toeing the establishment line.

While we will discuss the history of ESEA activism and community organising in another chapter (see Chapter 8: Activism and ESEA communities, historically), for now, we return to the sentiment of James Watson's statement at the start of this chapter: to migrate is to become a moving target. ESEA peoples and communities have been here for a very long time, and their struggles are not new – rather, they are in a constant state of flux, with cumulative impact on subsequent generations. If we are to move towards greater community cohesion and, indeed, move under a political identity, we must start by uncovering hitherto little known fragments of histories, and stitching them together.

Communities – fractured, imagined or otherwise – will always shift with migration in response to global and local shifts in climate, policy and economy. It's hard to imagine the future of ESEA communities against the future threat of a society increasingly divided over class and racial lines, the grip of right-wing populism and climate crisis never far from sight, but many of the constructed communities and spaces we have mentioned are already changing rapidly.

The family run restaurants, shops and takeaways that have shaped much of our understanding of ESEA migrant communities are starting to decline, with parents selling their businesses on retirement. Most had never intended for their children to inherit the business, and wanted them to have more opportunities for study and work (Ma, 2022). Meanwhile, the NHS is facing a 'brain drain' crisis, with high levels of resignation as workers either change careers or leave for more rewarding shores (Leary et al., 2024). With more remote jobs and new modes of working, ESEA people are more widespread throughout the UK and the ethnic minority population of historically white majority cities is increasing. The mixed population is also one of the country's most rapidly growing ethnic groups, as we'll discuss later (see Chapter 5: Common assumptions about mixed people).

While we address the future of advocacy at a later point within this book (see Chapter 8: The future), we hope for now we have gone some way to illustrate the violence of borders, and that the struggles created by punitive immigration policies foreshadow any meaningful and creative exploration of what it means to be a migrant in this country. For the simple act of being a migrant requires so much energy merely to exist within our borders. Even those who are granted status are faced with the constant emotional labour of complex housing bureaucracy, concerns over meeting minimum salary requirements, humiliatingly rigorous screening processes and complex and costly renewals processes. Additionally, they face feeling unable to participate publicly in solidarity movements or speak up against systemic racism out of fear for their migration status.

Borders delineate between communities and reinforce imagined or perceived differences, even between ESEA people themselves, and deprive our communities of the richness, solidarity and community connections they deserve. Is the pursuit of institutional inclusion besides the point, when the structures we currently exist under are geared towards exploitation along dividing lines of race, gender, religion, citizenship and class?

4

REPRESENTATION AND THE ILLUSION OF PROGRESS

THE LIMITS OF REPRESENTATION

Discussions and critiques on the much-debated topic of representation can often revolve around depictions in popular media, political representation, or the formation of leadership teams in professional organisations. The opportunities offered by white-led institutions are simultaneously celebrated and berated by ESEA audience members who seek 'fairer representation'. Scarcity of roles can leave us battling over crumbs and questioning the legitimacy of our inclusion. It also leaves us asking what 'fair' really looks like.

Representation (in its simplest form) does not suffice, especially when it lacks depth and genuine curiosity. This can lead to shallow views where work is reduced to racial identity alone. However, we can't seem to get away from the subject. In a survey by ESEA Music, a non-profit founded by UK-based ESEA music industry professionals, on the experiences of ESEA artists

in the UK music industry, one response highlighted the insepa-
rability of race, ethnicity and representation:

> I remember seeing a review of one of my albums criticise me
> for something along the lines of 'not using more elements of
> his Indonesian background in his music', whilst another review
> of the exact same album praised me for something along the
> lines of 'including elements of his Indonesian background in
> his music', despite the fact that as far as I was aware, I hadn't.
> Both were frustrating as it felt like, by virtue of being half-
> Indonesian, anything I did was typecast into what my racial
> identity was – like I was EXPECTED to only create something
> that had some relation to my race, or else get marked down.
> (ESEA Music, 2023)

This response underscores the frustration and limitations which
ESEA artists can face when their creative output is pigeonholed
into racial expectations, rather than being appreciated on its
own merits.

The lack of opportunities to tell meaningful stories and steer
decision-making conversations is a difficulty faced by many
ESEA people working in various industries and initiatives. Cries of
#RepresentationMatters and #YouCannotBeWhatYouCannotSee
can miss important discussions about *whom* it matters to, to
what end this representation serves and how it excludes people
and experiences that are not 'represented'.

In a tweet, Dr. Kelebogile Zvobgo, founder and director of
The International Justice Lab, states, 'Motion to replace "underrepresented" with "historically excluded." Precision matters;
the former is a consequence of the latter. Let's not forget'
(Feminism Fierce, 2021).

Zvobgo observes the impact of intentional acts of exclusion that has led to the modes of 'representation' we see in

current day contexts. In this chapter, we will look at ways in which organisations and institutions – under the guise of antiracism and 'inclusivity' – do little to materially help those they are seeking to elevate, instead leveraging 'diversity, equity and inclusion' initiatives to control the brand image. In another example, their complicity in regimes that have broken international law calls into question how ESEA organisations can find themselves the palatable face of an ethically compromised brand.

It's not that representation isn't important; indeed, the foundations of besea.n were built on this concept of broader representation, where we drew a line from the unnecessary use of ESEA people in COVID-19 news headlines to the real life impact on individuals targeted by anti-ESEA racism. While this campaign was the springboard for our platform, we have found – through our subsequent work and engagement with a range of organisations and institutions – various ways in which representation is being misused in contexts that can appropriate the 'ESEA' social construct.

SHIFTING INTENT TOWARDS IMPACT

Representation, if anything, is only a starting point, not the destination. Industries and the organisations within them are now faced with the necessary work that must follow beyond representation through to accountability – both on the part of those given the opportunity to represent and by those granting it. Representation without responsibility and discernment risks being tokenistic rather than transformative.

In short, it's important to consider what representation actually achieves – and to acknowledge where its limits lie.

There is no denying that the achievements by ESEA figures in various fields lead to elation for many ESEA communities and beyond. In sports, we witnessed this in the overwhelming national pride following Emma Raducanu's maiden Grand Slam Victory in 2021 and, further afield, Sunisa Lee's back-to-back Olympic medals. On our screens, we saw drag superstar Nymphia Wind being crowned the Season 16 winner of *Rupaul's Drag Race* in 2024 after proudly showcasing Taiwanese culture throughout the competition, and the record-breaking success of television series *Shōgun* winning 18 awards for its first season. Many celebrated Michelle Yeoh and Ke Huy Quan's historic Oscar wins for the 2022 indie film turned crossover hit *Everything Everywhere All At Once.*

The joy of seeing faces like ours on screen is only fleeting without meaningful change in our immediate realities. Beyond idealising success as a distant mirage in popular media, we must shift our focus closer – to the contributions and concerns of those most marginalised, whose calls for systemic change demand urgent attention. This serves as a reminder that increased visibility does not always translate to greater significance.

Consider the Vietnamese, Cambodian and Rohingya refugees in the UK who face systemic barriers in housing, employment and legal status (Barber, 2018), or the undocumented Burmese and Thai migrant workers at risk of wage theft and labour exploitation (Scott, 2017). Reflect on the double marginalisation faced by groups such as the Karen, Hmong and Cham, who are even more vulnerable to legal precarity – having faced histories of oppression in their home countries, which can influence how they are perceived and treated in the

UK (Balčaitė, 2019).[1] We must also acknowledge the struggles of low-income ESEA families, queer and trans individuals, and the elderly and isolated – particularly those experiencing language barriers. Additionally, the ongoing stigma surrounding disability and mental health continues to hinder neurodivergent and disabled ESEA individuals from accessing culturally competent support.

THE UNSTOPPABLES: MIGRANT DOMESTIC WORKERS SPEAK OUT

2025 saw a public showcase presented by The Voice of Domestic Workers, an education and campaigning group that advocates for the justice and rights of Britain's migrant domestic workers, founded by Marissa Begonia, a former migrant domestic worker herself. Participants shared their experiences of being trapped in abusive contracts, having their passports withheld, and being denied access to private bedrooms, in addition to many other intolerable conditions they have been subjected to. Since the UK government removed workers' rights and protections under the Overseas Domestic Worker

---

[1]The Karen people have faced decades of persecution by the Myanmar military, and upon resettling in the UK, they can encounter prejudice tied to their refugee status. The Hmong, originally from Southeast Asia, particularly Laos, were often viewed with suspicion due to their association with the US-backed forces during the Vietnam War, leading to discrimination in resettlement countries. The Cham people, primarily Muslim, have faced both ethnic and religious discrimination, particularly after fleeing persecution in Southeast Asia.

visa in 2012, they have consistently protested these changes, which have left workers vulnerable to exploitation.

At the showcase, one speaker named Christine shared her experience as a domestic worker, recalling times she went to work hungry because she was denied regular meals. Others recounted incidents where the parents of the children they cared for encouraged mistreatment. One worker described how she discovered The Voice of Domestic Workers on Facebook; after sharing her situation with them, she was rescued in the middle of the night.

As these stories were told one after another, directly from those who had endured them, denial became impossible. The horrific conditions domestic workers and others face are not fiction but reality. Platforming these stories is a vital step toward driving meaningful change, recalling multiple rungs on Yujie Zhu's heritage interpretation ladder (see Chapter 2: A purpose beyond a label): knowledge and truth-telling; learning and understanding; imagination and immersion.

The showcase demonstrated why representation must work at a grassroots level. This means ensuring the power to shape and share solutions, policies and narratives resides within those directly impacted, and that the efforts are sustainable. When we preoccupy ourselves with visibility on its own without materially addressing the deeper structural barriers affecting those most marginalised, even the most nuanced representation will remain an illusion of progress.

The horizon of change lies in tackling state violence and dismantling the carceral system that perpetuates these barriers. Racial profiling, lack of mental healthcare and low income making adequate legal support inaccessible, are a few of the reasons why marginalised people are targeted and over-policed in the prison industrial complex (Hyunh and Wemyss, 2021). Instead of people-centred rehabilitation, healing and restorative justice, current carceral systems exacerbate cycles of oppression. Initiatives like *On Your Side*, a specialist support and reporting service for ESEA communities, provide essential resources for those who have experienced racism or hate. However, while it may bring a specific type of public recognition about the severity of ESEA hate crimes in the UK, it also inadvertently reinforces a preference for more policing as a means of making communities safer.

Diasporic artist and writer Pear Nuallak (2023) aptly observes the tendency in most representational efforts stemming from a desire 'to be included within, rather than abolish, the carceral-colonial status quo.' Visibility and inclusion alone do not guarantee tangible impact or systemic change. Liberation begins when we stop seeking acceptance within broken systems and instead focus on dismantling them – ensuring that those who have lived the experience are the ones shaping the solutions (hooks, 1994).

For now, we'll put it simply: let's stop finding ways to sit at tables we are meant to flip.

RECOGNITION WITHOUT REDISTRIBUTION: WHO TRULY BENEFITS?

In January 2023, *Town & Country* magazine reported on an event hosted by Buckingham Palace, which was the setting for 'a reception to celebrate British East and Southeast Asian communities, and guests included representatives across armed forces, the arts, fashion, business, government and many more fields' (Burack, 2023). While the event aimed to 'celebrate' ESEA communities, we had to wonder what material support the Royal household extends towards the wider population who are excluded from these circles.

Meanwhile, King Charles increased his wealth by £10 million in 2024, according to *Time Magazine* (Gordon, 2024), bringing his net worth to £610 million. He also avoided paying any inheritance tax on the vast fortune he inherited from his late mother. Furthermore, the monarchy is exempt from Freedom of Information Requests, which shields them from any type of accountability that you would expect most public institutions to be held to (House of Commons Library, 2021).

Amid reductions to the winter fuel allowance and a cost of living crisis, the cost to the taxpayer for funding the monarchy's activities will increase by £45 million in 2025–26, reaching £132 million (Lavelle, 2024). While participation in the Royal event dedicated to select ESEA people offered a temporary invitation to rub shoulders with an exclusive set, ESEA organisations and individuals should be wary of how their presence is a beneficial PR exercise for the Royal household, with little material change in the livelihoods of those most in need. If we are working towards equity, celebration can only be found when we hold unfair systems accountable as one of the many and certainly most visible gatekeepers to the resources so desperately needed by the people they extort.

Too often, institutions embrace diversity in a symbolic sense but fail to stand firm when that representation is challenged. The absence of preparedness to support and protect the very people they choose to feature reveals a deeper issue of representation being treated as an aesthetic choice rather than a meaningful commitment. This reluctance is itself a form of privilege; where organisations can pick and choose when to engage, leaving marginalised communities to bear the weight of backlash alone.

In May 2024, a National Trust social post featuring Asian people was met with hostile comments demanding justification for their presence in the organisation's marketing, claiming the organisation 'don't want whites' or don't use enough white people in their advertising (National Trust, 2024). Incidents like these highlight the need for organisations to stand by their commitments to diversity – not just in their imagery but in actively addressing and challenging racism and how marginalised people face barriers to access National Trust spaces. This includes formulating clear strategies to protect both their digital and physical spaces from identity-based harassment, educating audiences on the importance of inclusive representation and spearheading initiatives to make the natural spaces more accessible to the most marginalised.

We can observe the same extractive mindset as we consider institutions through the same lens, where entitlement and exploitation by white-led institutions manifest in more insidious ways. It is no longer just land or labour being taken, but intellectual property, where the work of already marginalised individuals is co-opted and erased. The irony is especially stark when institutions with histories of cultural theft continue to engage in intellectual exploitation.

One such example occurred in June 2023 when the British Museum plagiarised translator Yilin Wang's work for its *China's Hidden Century* exhibition. Her translations of Chinese revolutionary Qiu Jin's poetry were used without credit, consultation, or compensation. Only after she publicly called out the museum on social media did they acknowledge the oversight and belatedly offer remuneration. The incident was later dismissed as 'unintentional' (Khomami & Hawkins, 2023) by one of the exhibition's principal researchers, Judith Lovell, professor of modern Chinese history and literature at Birkbeck

University, exemplifying how institutions often fail to audit their own practices while claiming to champion inclusivity.

Here, we see an example of representation where inclusion is performative rather than transformative, where institutions claim to celebrate diverse voices while simultaneously erasing or exploiting the very people they seek to uplift. This pattern is not unique to the British Museum but reflects a broader institutional habit of extracting labour, knowledge and culture from marginalised communities without accountability. We explore patterns of exploitation, appropriation and workers' rights in Chapter 6.

Efforts for equity in the workplace still hold relevance, we only hope not as an HR function that primarily serves the business or organisation, but as a structural necessity in navigating power dynamics to protect workers and ensure transparency for consumers. After all, these initiatives have their roots in civil rights movements and union organising, to secure hard fought workers rights for women, racialised, disabled and LGBTQ+ people. Representation must go beyond surface-level inclusion and into real institutional accountability, ensuring that when marginalised communities are visible, they are also defended – and not just in the Global North.

Writers like Ash Sarkar have highlighted the absurdity of global corporations like Lockheed Martin and Coca Cola publicly celebrating their involvement in DEI initiatives, when they are responsible for moral failures like missile systems used in illegal wars, labour exploitation and resource monopolisation in the Global South, to name just a few human rights violations (Sarkar, 2025). Representation cannot be reduced to visibility alone; it must come with respect, credit and material redistribution. Too often, the irony of representational 'visibility' is that it has become a handy tool for big corporates to cover up their most egregious crimes.

RETHINKING SPONSORSHIP AND FUNDING: CULTURAL BOYCOTTS AGAINST COMPLICITY

In 2024, besea.n secured a sponsorship opportunity for our flagship ESEA Heritage Month event from a large food manufacturing company. Their product range includes frozen food and seasonings, with monosodium glutamate (MSG) being the most popular. The brand has campaigned hard to debunk the racist myths surrounding MSG. In 1968, a letter published in *The New England Journal of Medicine* described symptoms of weakness, numbness and a racing heart after the writer consumed Chinese food, suggesting that MSG[2] may have been the cause. Several robust scientific studies have since debunked these allegations and the brand has attracted several celebrity supporters who have worked to rehabilitate the image of MSG.

besea.n has no reliable means of income and is reliant on the generosity of important but relatively small monthly donations, so the chance to partner with a family-favourite brand that invests in anti-racism initiatives felt meaningful and relevant. When news of our sponsorship was shared on our social channels, we were informed of the brand's investments in Israeli companies. The Boycott, Divest and Sanctions (BDS) campaign

---

[2]MSG is a flavour enhancer developed by Japanese chemist Kikunae Ikeda, who isolated the chemicals naturally found in food responsible for eliciting the umami flavor. Its production and use has been mired in racism for decades.

calls for boycott and divestment from all Israeli companies unless they meet conditions of non-participation in Israeli apartheid, settler colonialism and military occupation, and the public recognition of Palestinian rights under international law (Barghouti, 2011). According to the Palestinian-led movement, which was inspired by the South African anti-apartheid campaign, no single Israeli company meets these two conditions. besea.n supports the struggle for Palestinian freedom and used our platform to call for the disarmament of the Israeli military, so it was a quick decision for us to withdraw from the partnership and return all the funds donated by the brand.

Cultural boycotts are a significant tool in the fight for freedom against organisations that benefit from settler colonialism and the violent extraction of resources in occupied territories. A major walkout by authors and disruption from activists effectively ended the 20-year partnership between the Edinburgh International Book Festival and the investment firm Baillie Gifford over its links to fossil fuels (*The Guardian*, 2024a). While many lamented Baillie Gifford's decision to simply stop funding literary festivals rather than focus on full divestment, the action highlighted the unsustainability of arts funding mechanisms that rely on the support of problematic funders. Similarly, the group Bands Boycott staged a campaign to boycott music festivals sponsored by Barclays Bank, which provides financial services to, and manages client shares in, companies implicated in Israel's military occupation. In 2024, public pressure combined with a walkout by artists ended the affiliation between Barclays and The Great Escape Festival (Richards, 2024).

The influence of investment companies in the cultural sector is extensive. While Barclays' sponsorship of various music events has since been suspended, other financial giants continue to play a significant role. Aviva, the UK's largest insurance and pensions provider, invests in companies that supply arms to Israel, which have been linked to the oppression of Palestinians (*Canary*, 2025). Aviva Studios is based in Factory International, the organisation behind the Manchester International Festival which also funds a year-round programme of creative events. This is just a glimpse of how deeply everyday consumers are entangled in a system that prioritises wealth accumulation for a small group of billionaires.

Arts funding in the UK has faced significant cuts in recent years. Between 2010 and 2023, funding for cultural institutions fell by 18%. From 2023 to 2024, 32% of museums experienced reduced or entirely withdrawn funding from local authorities. Arts Council England announced £56 million in cuts to arts funding in London from 2025 to 2028 (House of Lords Library, 2024).

These severe funding cuts have left cultural groups and local community initiatives with few options for survival. The Oxfordshire Chinese Community & Advice Centre and Wai Yin Society in Manchester both support hubs for local ESEA communities and have faced financial struggles despite providing important culturally-aware services. The An Viet Foundation, set up in 1981 after the Second Indochina War, became a central support hub for resettled Vietnamese families in London as well as housing important archival documents. The centre closed in 2017 and An Viet Archives, a collective set up in 2023, are working to preserve the huge collection of historical records left in its wake. The closure of The Vietnamese Mental Health Services, along with China Exchange, an organisation that had advocated for heritage education in London's Chinatown for 9.5 years, further marks the stark reduction of community centres relied on by so many.

Grant opportunities are scarce, and the application processes are often intensive, creating barriers for volunteers and those with limited experience or time to complete them. As a result, smaller grassroots collectives are often left with no choice but to rely on sponsored events from larger corporations that may profit from exploitative practices.

REFRAMING REPRESENTATION TO BRIDGE EMPATHY AND FOSTER AGENCY

Beyond institutional changes, we must question our desire for acknowledgment. Why continue seeking validation that conforms to colonial expectations? The obsession with 'good' ESEA representation often leads to a narrow view of success, one that panders to sanitised, palatable stereotypes rather than embracing the full complexity of people in ESEA communities. We must shift from seeking approval based on idealised narratives, and instead break free from eurocentric standards. Progress should be defined on our terms, grounded in our experiences and free from mainstream media filtering. It means looking at Britain through our own lens, not the other way around.

Perhaps we can reframe representation as a tool that encourages wild permission, offering the freedom to reimagine how we can be and to develop our own metrics for success, instead of curating a retrospective version of how we think we should be. The American actress of Chinese heritage, Joan Chen, echoes this with a call towards reclaiming our own agency in narratives instead of relying on what is presented to us:

> Stereotypes are not necessarily untrue or malicious, though they can be as malignant as cancer. Their danger lies in their

one-dimensional quality, its incompleteness however no one can fill in the blanks. No one can complete the stories for us. Only we can. And the best way we can repudiate a single story told about us by the dominant culture is to tell 1001 of our own tales. (Committee of 100, 2017)

It is through the abundance and diversity of nuanced ESEA narratives that the stronghold of harmful, monolithic stereotypes will dissolve. When the general understanding of ESEA people is no longer limited to the few stories typically portrayed, we can create space for more versions of ESEA stories to emerge, free from standards set by the colonial gaze. In other words, narratives that reflect the reality, complexity and humanity of ESEA experiences, even if they don't always conform to the mould of traditional 'success' stories. It's less about the scope of visibility and more about ensuring there is space for expression, at the very least, to practise taking up space. To practise courage to engage in the first place rather than to default to silence.

Representation, when approached as a practice in this way, can be empowering. It fosters agency by reminding individuals that their voices deserve space – taking the very first step away from feelings of helplessness and insignificance. Beyond this, a growing curiosity and a genuine interest in storytelling can elevate representation into a social practice that bridges empathy and diversifies narratives. We recognise such efforts of fostering agency and curiosity in kindredpacket's 2021 campaign, *Stories of Our Heritage*, an open call for the wider ESEA community to submit personal stories about living at the intersection of British and ESEA experiences. By engaging with each other's narratives, individuals deepen their understanding of their own history – through remembering and imagining together.

Along similar lines, representation can also be utilised at a grassroots level to enable like-minded people to find each other in the practice of reclaiming physical spaces. ESEA Outdoors, a community group founded by activist Andrew Wang, harnesses the power of social media to connect ESEA individuals across the country through shared outdoor experiences, featuring a map that anonymously shares members' geographic locations. In doing so, the map also generates visibility of ESEA people in outdoor spaces in hopes of affirming their presence and rightful sense of belonging, while reducing the risk of hostility from others to promote safer access to nature and the outdoors.

Calls for representation also present an opportunity to bridge empathy and connection. Nature, language and food in particular, serve as powerful mediums for representation, transcending political and religious differences while seamlessly integrating into daily life. They allow more humanised exchanges between communities, making empathy and connection more accessible. Reverend Mark Nam, a Berkshire-based Vicar, captures this through his work with The Teahouse, an apolitical support network for Chinese-heritage clergy:

> In serving tea, we're not just offering a drink, but a moment of shared culture and understanding, a step towards bridging our differences. (Nam, 2023)

ESEA Green Lions' *Microseasons* project invites ESEA individuals to contribute to a year-long community piece, using observation and interaction with the natural world, weather and climate around us – or at least creating a relationship with the environment. Similarly in 2024, #AsianDessertExchange, a baking collective founded by author and community organiser Jenny

Lau, offers a playful yet meaningful approach to intercommunity exchange through their bake sale *Sugar Spice Bread Rice*, by pairing bakers from across the Asian spectrum to collaboratively explore new flavours through the common language of these basic ingredients.

BEYOND VISIBILITY AND INTO ACTION

In recent years, inclusion has become a buzzword that is worn like a badge, where little substance presides behind its veneer and its very existence as a concept is proof that division and hierarchy continues to exist.

Representation, in its most radical form, should not serve as a means to simply populate existing structures with more diverse faces, nor should it be reduced to a tool for personal gain or institutional capital. Instead, it should be leveraged to build solidarity, empower grassroots movements and create tangible pathways toward collective liberation. At its core, representation should function as a bridge, enabling like-minded individuals to find each other, forge strong networks and fight against shared struggles. Rather than seeking approval within dominant systems, we must use representation as a way to strengthen intercommunity relations – recognising that our struggles are interconnected and that real power lies in coalition-building. This means moving beyond siloed identity politics and toward cross-community collaboration, learning from the histories of other marginalised groups, and working in solidarity toward systemic change.

The ability to tell our own stories has never been more accessible. From the early days of LiveJournal and Tumblr to the expansive reach of Instagram and TikTok, social networks have allowed individuals to bypass traditional gatekeepers and broadcast their own narratives. The challenge now is to resist the pressures of external validation and instead use these platforms to shape discourse on our own terms. Representation should not be about centring our ESEA identities as the defining aspect of our work, but about recognising how our shared connections can be mobilised for a greater purpose. By expanding the scope of representation beyond visibility and into action, we can collectively reimagine what solidarity looks like – one rooted in agency, intercommunity support and a refusal to be defined by the gaze of others.

5

MULTICULTURALISM AND MIXED ESEA EXPERIENCES

MIXEDNESS AND THE LANGUAGE OF FRACTIONS

There is an unsettling dichotomy that often comes with being a person of mixed heritage. While many mixed people feel these days they can be openly proud about their mixedness, our society has an overwhelming preoccupation with fractions, which can encourage feelings of inadequacy. When we reduce identity to percentages, we assume that genetic makeup alone prescribes one's ethnic or cultural identity. There is a fine line between analysing genetic make-up and what is essentially validation by numbers: 'Yes, she's Japanese, but **only** half' or 'He's fully Chinese **but** doesn't speak the language.' According to this logic, to be monoethnic means you check all the boxes and cultural benchmarks we impose on a particular group (language, physical appearance, practices, beliefs, and so on). By contrast, then, to be mixed implies inadequacy in one or more areas. Sadly, refrains like this are commonly spoken by people of all ethnicities, including among ESEA people and, unfortunately, mixed people themselves.

Among Asian communities in diaspora, there has been some attempt in the last few decades to co-opt language for meaningful mixed Asian identities, such as the Hawaiian and Japanese derived terms 'hapa' and 'haafu' (Parker & Song, 2001). It also must be said that many mixed people *are* comfortable with fractional identifiers. While we encourage adoption and reappropriation of language where the outcome is empowerment and reclamation, we should be wary of the terms' insistence of fractional representation of mixed people, since they may, 'in a race-conscious society, serve to reinforce the ideology that the mixed race individual is somehow less than a whole person' (Parker & Song, 2001). To tackle the exclusion that these potentially divisive terms imply, we need to do much more than change our language.

In the UK, there is a tendency for people to assume by default that 'mixed race' is code for 'mixed Black and white', which not only centres whiteness, but contributes to the erasure of other mixed experiences. Terms such as 'mixed' or 'mixed ethnicity' are coming to the fore because they are more inclusive of those people with two Asian biological parents of different ethnicities. As with most things related to identity, however, it's not just as simple as there being a right or a wrong way to identify: mixedness is complex and nuanced, and the simplified language of binaries will always fall short when we attempt to make it serve something so multi-faceted. Despite this, some people may identify as 'mixed race', and others won't, so we encourage everyone to consider their own situation and which language they feel suits them best.

Since the introduction of the 'Mixed' category for ethnicity in the Census for England and Wales in 2001, there have been increasing conversations and studies on the growing demographic. For those of us who grew up in the 90s and the early 00s, mixed relationships were more accepted than they were among the previous generation. However, there were limited in-depth conversations on mixed identity among multiethnic households, which may simply have been a question of the language that was available at the time. Those born before 2001 would have had no way to record their ethnicities accurately in official data collection. As Natalie Morris writes in her book *Mixed/Other*, 'An entire mixed generation grew up watching the world define and categorise our identity in real time' (Morris, 2022).

Despite our reluctance to speak prescriptively about mixed experiences, and in recognition of the importance of affording mixed people the agency and intention with which to undertake and define their own identities, it would be remiss of us not to mention imposter syndrome and the feeling of non-belonging that many mixed people *do* feel. Making space for mixed experiences means understanding and welcoming both positive and negative narratives: from embracing a rich cultural heritage, to feeling excluded in certain spaces and suffering a particular brand of anti-mixed racism. For many, going through life as a mixed person means an existence where it's open season on your genetic makeup. It is a particularly demoralising experience to have the different aspects of your face, hair and skin appraised and critiqued by total strangers, some of whom are trying to decide which ethnicity shines through the most, and some of whom are simply admiring your 'exotic mix'.

Writing for besea.n, Tomi Haffety, who claims both white and Japanese heritage, felt encouraged by the #StopAsianHate movement of 2020. She was stopped in her tracks, however, by imposter syndrome, wondering if she was 'Asian enough' to get involved. These feelings were strongly influenced by her family and the conservative views of wider Japanese society towards non-ethnically Japanese people, who have always told her she does not look Asian at all.

While Tomi makes efforts to explore Japanese culture, she recognises that such effects may be an attempt to 'prove to other Japanese, and Asian people, that I have reason to be given the label when I feel so culturally British' (Haffety, 2021).

The intentionality behind these efforts to find belonging and assert identity, or the idea of having to 'work' to fit in, are very common among mixed people. Writer Afua Hirsch, who is of mixed white British, German Jewish and Black Ghanaian heritage and grew up in Wimbledon – an affluent, largely white and middle-class area of London – speaks of the collision of her ideas and experiences with those of her partner, a Black British man from a Ghanaian family based in Tottenham – by contrast, one of London's poorest and most multiethnic neighbourhoods. While she was afforded privilege from her culturally white British upbringing and environment, which allows her, as a mixed woman, to move in certain spaces with more palatability than a dark-skinned Black man, Hirsch admits that she felt envious of her partner's exposure to, and immersion in, Ghanaian cultural identity, which she felt he had absorbed 'by osmosis' (Hirsch, 2018).

For some, moving alongside those who have never questioned their identity can be a difficult feeling to reconcile, as Tomi also observed, admitting that the lack of relatability was a lonely experience. What we increasingly discover, when we make space for a multitude of voices, is that there is no one way to be anything. There is no singular British Ghanaian experience, nor British Japanese. It is also by the grace and wisdom of Black communities that ESEA people can learn to make space for mixed experiences. In recent years, writers and thinkers in Black cultural identity spaces have called for greater inclusion of mixed Black people. Despite having some non-Black heritage, they will still be Othered in society, and therefore must be welcomed with open arms into Black communities. Of course, cultural gatekeeping and marginalisation does still occur, which is why we should always approach inclusion with diligence and awareness.

COMMON ASSUMPTIONS ABOUT MIXED PEOPLE

It's important to acknowledge that many of the identity struggles experienced by mixed people also impact those who grew up with two parents of the same ethnicity, but in a different country to their parents' country of origin. There may be cultural misunderstandings, differing attitudes towards collectivism versus individualism, or even a language barrier. However, mixed people often face additional challenges – and experience privileges – specifically *because* of their mixed heritage. Mixed ethnicity people are one of the UK's most rapidly expanding ethnic demographics, skyrocketing from 677,177 in 2001 to 1.7 million in 2021 (Diversity UK, 2023).

Mixed experiences are infinitely diverse and wide-ranging, especially since a greater range of ethnic makeup provides greater potential for intersectionality, making it impossible to condense all mixed experiences into one set of commonalities and themes. Throughout this chapter, we've tried to bring in a diverse range of voices who can speak to different aspects of mixed experience and multiethnic multiculturalism. We encourage you not to view these experiences as representatives of certain communities, but rather, as single pockets of insight among a rich and vibrant chorus of ESEA voices.

Many mixed people will be familiar with the minor existential crisis that is invoked by the exasperatingly inconsistent categories for ethnicity that plague most form fills in the UK (see Chapter 3: ESEA people in the UK today: on paper). When 'Mixed Asian and White' undoubtedly refers to *South* Asian and white, selecting 'Mixed Asian – Other' is a literal exercise in self Othering. However, it is commonly assumed that all mixed people live in a perpetual state of confusion and existentialist angst.

Professor Miri Song introduces the concept of 'ethnic options' and the choice or ability that people have to 'exercise' these options. She writes:

> … to possess ethnic options in any real sense must mean that groups are actually able to claim at least some images and identities they desire, in a variety of social contexts and especially in public spaces. In other words, the ethnic identities asserted by groups need to be recognized [sic] and validated by the wider society – otherwise, the term 'ethnic options' carries no real weight or significance. (Song, 2003)

The availability of and access to these 'ethnic options' depends largely on the different types of privilege afforded to an individual (see Chapter 2: Language, citizenship and the politics of belonging). Access to social, economic or cultural capital specifically influence a person's ability to exercise their 'ethnic options' – simply put, it's easier to feel comfortable with your ethnic and cultural identity *and* your place within our current social system the more capital you have.

So how does access to capital and ethnic options impact mixed people? We interviewed a number of mixed ESEA people with this in mind, and it was clear that all the interviewees were highly impacted by the environment they grew up in, the countries they lived in, the experiences they had both within their families and outside their families, and access to affirming spaces with other mixed people.

A description commonly ascribed to mixed people with white heritage, who have features that are *perceived* to be predominantly white, is the phrase 'white-passing'. A number of people we spoke to indicated that this phrase made them feel uncomfortable, as though a) all mixed white people automatically have aspirations to 'pass' as white or b) their 'whiteness' is their defining characteristic, invalidating all other traits that compound to build cultural or ethnic identity. Ironically, whiteness in mixed people is often the trait that white people themselves pay attention to the least. While we can and should acknowledge that having lighter skin or eurocentric features often affords a level of privilege and shields against the most obvious types of racism and discrimination, the term can be problematic.

Referencing a study of interracial Japanese Americans (Mass, 1992), Miri Song explains that mixed people living in majority white societies may have an even stronger sense of identity with regards to their racialised ethnicity due to being subject to greater criticism and pressures of belonging. As a result, they may make more conscientious efforts to embrace and celebrate their ethnic identity. Finally, in reference to the same study of Japanese Americans, Song points out:

> The fact that many mixed individuals of White [sic] and Japanese heritage look more White than monoracial Japanese does not automatically ensure that they will *feel* less Japanese. Feeling Japanese, however, does not ensure recognition of their Japanese identities by other (monoracial) Japanese people or the wider society. (Song, 2003)

Translation? People often treat mixed people based on how they look on the outside rather than what they feel inside. We must, therefore, strive for culturally inclusive spaces that do not make judgements or assumptions of a common identity, but rather, which invite curiosity and difference in the hope of forging new connections. It is also vitally important that we acknowledge the consequences of centring whiteness and eurocentric features on wellbeing and body image and their impact on beauty standards in ESEA countries and in diaspora. In addition to the mass-marketing of dangerous skin lightening creams across Asian markets, ESEA women are increasingly opting for cosmetic procedures such as eyelid surgery or so-called 'Asian rhinoplasties', of which 650,000 were performed in South Korea in 2011, making it the country with the highest number of procedures in Asia (Asian Century Institute, 2014). Writing on the topic for the *Huffington Post*, Adora Svitak said:

> An entire clinic in Southern California, the 'Asian Eyelid Center' exists for the specific purpose of surgically creating creases. Never in my childhood had I thought that a tiny fold of skin would be part of my privilege. (Svitak, 2014)

Svitak goes on to explain the Asian/Caucasian beauty dichotomy created by the centring of whiteness and the hyperfetishisation of the exotic East: '[the] goal is to be simultaneously the perfect Asian and the perfect Caucasian – a china doll, with big eyes and eyelid creases.'

It may be that mixed ESEA people with white heritage are perceived to be closer to achieving this aesthetic goal, but we must actively reject its glorification among our communities – firstly, because it erases and invalidates those with darker skin or certain more typically 'ESEA' facial features, and secondly, because it teaches mixed people that their value is rooted solely in their appearance – something beyond their control, assigned by the lottery of birth.

TIME, PLACE AND SPACE

In a podcast interview (Vu Peterson, 2021), Mai-Anh spoke with Alicia Warner, photographer and illustrator of Thai and Antiguan heritage, from London, and Francesca Humi, community organiser and cross-border activism coordinator of Filipino, Italian, Austrian, Hungarian, Jewish heritage, from Paris. They both acknowledged that they 'choose' how they decide to present their identities depending on how they feel, who is asking and how the situation feels to them. Francesca recalls discussing with her parents how to identify herself, admitting that memorising the long list of identities was quite a challenge for a 7-year-old. Often mistaken as 'la chinoise' – 'the Chinese girl' – growing up in Paris, it was only after moving to Montreal, which has a strong Asian population, that she realised it was possible to identify *as* mixed.

Being surrounded by a variety of different Asian nuanced identities allowed Francesca to recognise the racialised part of her identity more intentionally, and prompted her to connect more strongly with her Filipino heritage. By contrast, moving to London and finding that Filipinos are relatively under-acknowledged only strengthened Francesca's resolve to advocate for and embrace Filipino communities. In Francesca's experience, many people she met were only really aware of Filipino communities in London in a healthcare context, Filipino being the second most common non-British nationality in the NHS (Baker, 2023), or as domestic workers or nannies.

Shifting identities in different environments is a very common phenomenon for mixed people, it seems. Alicia had comfortable access to a diverse group of friendships and spaces to explore in Southeast London as well as strong support from both sides of her family. These optimal conditions allowed her to create a strong foundation of mixed identity. In Thailand, however, her Thai identity became muddled, partly because of ignorance around Black people that inevitably preoccupied many Thai people she encountered, albeit largely through curiosity rather than hostility.

When Francesca moved to the Philippines, her relationship with her identity shifted again. 'To be a mestiza in the Philippines comes with a lot of privilege, and that is an extremely colonial identity,' she said, going on to explain the impact on current trends such as damaging beauty ideals rooted in colourism – the preference for lighter skin that is prevalent in many communities of the Global Majority. To some extent, in ESEA countries, a desire for lighter skin is often linked to an association between sun-darkened skin and the work of field labourers, traditionally of lower economic status.

However, it would not be unfounded to conclude that the colonisation projects of European powers in Asia simply built on an existing system of social injustice and colourism. This was strengthened by a structure based on racism and white European supremacy, especially in the Philippines where the Spanish regime imposed a strict class system based on ethnicity and skin colour (Albiez-Wieck, 2025).

TRANSRACIAL ADOPTION: A DIFFERENT KIND OF MULTICULTURAL UPBRINGING

While they will undoubtedly experience their own unique identity evolution relative to their particular circumstances, many transracial adoptees (children adopted by parents of a different racial background to them) also resonate with the mixed person struggle of having limited or differing access to certain parts of their cultural heritage in comparison to those who grew up with the same cultural influence on both sides. besea.n member Pippa Power, and besea.n volunteer Emily Couch, both adopted from the Philippines and China respectively, by white parents, told us about their experiences (Vu Peterson, 2022).

Both interviewees expressed the personal importance not only of open communication from their parents, but also of the regular presence of other friends and family from the same cultural and ethnic backgrounds. Both have siblings, also adopted, with the same ethnic heritage; Pippa grew up in a community of family friends who had also adopted transracially, and Emily has Chinese and mixed Chinese relatives in her immediate family. However, the familiar influence of place and space we discussed earlier has certainly played its part.

Pippa, who grew up in Hong Kong in an international community, explained that living in the UK has increased her feelings of displacement. She is more frequently treated as a source of knowledge on the Philippines (the country of her birth) and many people assume that she has the same cultural frames of reference as other Filipinos – as if, indeed, there is only one 'way' to be Filipino. Both Emily and Pippa shared feelings that may be familiar to many mixed people. These include: not having the language to discuss identity; downplaying differences in predominantly white spaces (which may have prevented a further exploration of heritage and identity); using humour as an easy way to navigate identity ('I used to tell people I was a banana', said Emily); and feeling imposter syndrome, especially in ESEA community spaces.

Emily believes that the fact that she doesn't speak any ESEA languages, or doesn't have the same cultural references as other Chinese people, made her hesitate to volunteer with besea.n: 'It felt like I wasn't like a real Asian person compared to other people, and I was worried that people would say I wasn't Asian enough to be here.'

Interestingly, imposter syndrome in social justice and community advocacy spaces may have the effect of the mixed person or adoptee downplaying or dismissing their experiences with discrimination because of the presence of other types of privilege. Emily spoke of the confusion between being unable to distance herself from her experiences as a racialised woman and being afforded the privileges of a white middle class upbringing. This wider point about privilege is likely to be the case for many people who aren't adopted, who also have access to education, generational wealth, citizenship and so on. While we shouldn't detract from the adoptee experience, these observations serve as a helpful reminder that we all benefit from interrogating our own relationships with race and class, but that access to those privileges should not prevent us from recognising the harms that may come our way because of our ethnicity.

Elsewhere in this book, we discussed the emergence and importance of the ESEA acronym as a political identification and emerging demographic (see Chapter 1: Language matters and Chapter 3: An ethnic identifier in the making). Though some may find it limiting, for others it may serve as a convenient alternative to the identities they have not quite gelled with in the past. For example, Emily does not identify as British Chinese, despite being British and ethnically Chinese; she wants to acknowledge her roots but doesn't want a label to assume a particular shared experience that she does not personally claim. In agreement, Pippa also emphasised that identifying as 'ESEA' is much more comfortable for her, thereby lending renewed importance to the emerging identities of the past few years and, indeed, of the need to make space for further emerging transracial adoptee identities.

Like Tomi and many other ESEA people, Pippa was confronted with the fears around the COVID-19 pandemic and how that impacted her feelings of identity – ultimately, the pandemic has forced many of us to reckon with what it means to be ESEA in this particular cultural moment. While Emily and Pippa both expressed gratitude for having parents who were consistently open and transparent about transracial adoption, both also admitted the limitations of such conversations, seeing as their parents will never fully understand their experiences.

Both Pippa and Emily confronted these mixed feelings by publishing personal essays on the besea.n platform. Pippa calls her article 'an outpouring of 30 years of emotions', but also admitted that she felt stressed about feeling the need to speak for an entire community. Eventually, she decided to speak for herself, finding that her attempt to publish a piece that would speak to the entire ESEA transracial adoptee community would actually lead, somewhat ironically, to self-censorship. There's something to be said for this tension of wanting to acknowledge the diversity of ESEA experiences without erasing your own. Perhaps such self-preservation is actually instrumental in creating shared experience: we are all different, yes, but our desire to embrace our differences and to forge spaces for our multitudes of identities actually creates a common bond.

There is no singular ESEA experience – and the fact that we're all different is what unites us all.

MAKING SPACE FOR MULTITUDES

As we've mentioned, the common assumption that 'mixed' always implies one white parent erases the experiences of other mixed people from mainstream social discourse. From our media to our politics, from our education system to our creative spaces, unless we make the conscious effort to centre a wide variety of different mixed and multicultural ESEA experiences – and move away from one or two binary modes of 'being' ESEA – we will never cross the threshold of true inclusivity.

In particular, due to the existence of anti-Blackness that still persists in many ESEA cultures, we must emphatically remind ourselves that being Black is not the antithesis of being Asian, as much as media-fuelled racial stereotyping would have us believe that it is. It is perfectly possible to be both Black *and* Asian, and to accept and embrace all that those identities bring to an individual. As our societies change and evolve, mixed ESEA groups are diversifying and new ways of being in the world are coming to fruition.

When Alicia spoke to us about how her Black identity showed up for friends and people in her entourage growing up, she shared anecdotes that may be familiar to many: friends commenting 'I always forget that you're Black', people analysing which of her physical attributes fit their expectations of Blackness or Asianness, or the same tired refrain about how interesting or exotic her particular heritage mix is (Vu Peterson, 2021). If reading these experiences makes you feel exasperated, you can imagine how it feels to live them on a daily basis. The takeaway we'll keep coming back to is simple: mixed people don't choose how their ethnicities will combine in any physical manifestation. Their bodies should not be up for discussion unless they are the leaders of that conversation – and this is a practice we should undertake for *any* person. It's just good manners.

We see colourism plainly in the preference for mixed (with white, of course), lighter-skinned actors and models. Since colourism is rooted in anti-Blackness, what does that mean for mixed Black and Asian people navigating the Asian parts of their heritage, particularly when they move in the spaces of their Asian parent's country of origin? For Alicia, the lack of awareness among native Thais of Black communities outside Africa strongly marked her experiences living in Bangkok after graduating university. Despite her multicultural London upbringing, when in Thailand, she admitted that she finds it hard to self-identify as Thai: 'I will never be perceived as Thai [there], so that's why I still carry that ";ลูกครึ่ง / luk khrueng' [half child]" card.' This realisation is not a self-defeatist one, however.

Alicia went on to stress that she believes all friction and conflicts of this nature stem from a lack of understanding: essentially, when we understand the root of something, we understand how to approach and tackle the issue. Additionally, the more that mixed Black and Asian people have access to spaces of exploration and space to embrace their identities, the more we will embrace what it means to be both Black *and* Thai – not a half, or a fragmented piece. Alicia is firmly decided on claiming a multitude of identities, insisting, 'I will just claim absolutely all of them. I'm quite happy to say I am a Brit, I'm Black, I'm half Antiguan, half Thai, that I also live in Scotland' (Warner, 2021a). She believes that change is coming in Thailand, with migration and the growth of Black Thai communities, and especially the digital age, with more Black Thai people influencing popular culture, though it has yet to spread to politics and institutions.

For Kyami Mitsui Russell (K. Russell, personal communication, 20 June 2023), a Manchester-based artist of Black and Japanese heritage, the identity enablement of shifting place and space also rings true. While they identify as a queer, Black and Japanese, femme-presenting person, their identity is also further complexified by questions of national identity. While Kyami grew up in the US, rather than the UK, they have spent most of their adult life in England, and explained: 'I've found the most connection to my heritage while I've lived [in England].'

British people often look to the US as having more established Asian American communities, but we overlook that not
everyone has access to the physical community spaces that
foster that sense of identity. Kyami likened identity to constructing a constant jigsaw puzzle – at times, you may find a
piece under the table and realise that a piece fits, but you're
still always building the picture. On heritage, they described
feelings of belonging to both communities and sometimes to
neither, and those feelings fluctuating over time. One thing
that felt consistent, however, was feeling pressure to be the
bridge between Asian and Black communities, and being
treated as a source of knowledge responsible for tackling
racism on both sides.

The problematic issue of whose job it is to educate the
uneducated is commonly felt in anti-racism movements. The
reality is that, especially with a wealth of information at our fin-
gertips – produced at great personal effort by writers, thinkers
and activists of the Global Majority – it should be everyone's
responsibility to challenge or interrogate problematic behaviour when it arises. While it may not be appropriate for people
who don't belong to a particular community to be at the
forefront of social movements that advocate for those communities, having conversations where we challenge certain
behaviours is definitely an act of solidarity that can, and should
be undertaken by anyone. To both Kyami's and Alicia's points,
no one knows everything, but approaching situations with
respect, grace and empathy is usually a good place to start.

The reality is this: not all mixed people will experience
an identity crisis, but many experience shifts in identity influenced heavily by their environments. If mixed people are
provided with safer spaces to explore their mixed heritage,

where diversity of identity and nuanced experience are championed, they can flourish. That's why we need both, to ensure that mixed ESEA people are welcomed into ESEA spaces, but also that they have spaces to explore specifically mixed identities. We cannot deny that friction between mixed-ESEA communities and non-mixed ESEA communities does exist, but it might be because we're all struggling for crumbs when we should have access to a whole buffet.

HUMAN COCKTAILS OF ETHNICITY

Increasingly, mixed identities are seen as cool and on trend, which, at first glance, doesn't seem like anything to be sceptical of. What's wrong with being seen as cool and exotic? However, like the model minority myth (see Chapter 1: Class and the model minority myth), belief in this phenomenon as a positive one risks totally overlooking, ignoring or outright erasing the specific struggles that mixed people may experience in society as a direct consequence of their racialised identities. In addition, as is often the case with the history of racialisation, there are some deeply unsettling reasons why the fascination with what Francesca calls 'human cocktails of ethnicity' (Vu Peterson, 2021) are problematic and dangerous.

Mixed Asian women are the most highly preferred (and therefore highly fetishised) dating group in American society according to a 2015 study on how multiracial daters fare in a mainstream online dating website (Curington et al., 2015). This has direct consequences for the physical and psychological safety of ESEA women as well as the implications for ESEA men (see Chapter 7: On desirability, fetishisation and self identity). It also can't be seen in a vacuum: just like the model minority myth, a valorisation of Asianness often requires the rejection and de-valorisation of Blackness. On dating apps, for example, Black people's messages are responded to the least out of all ethnic groups examined in a 2013 study on dating app preferences (King, 2013).

Our society's preoccupation with physical features in particular is disturbing; we must remember that not so long ago in the grand scheme of human history, people of the Global Majority were treated like animals for having physiological differences to white people. More recent examples of dehumanisation are seen in the mistreatment of mixed children in British society: while the UK never had anti-miscegenation practices enshrined in law like in the US, and despite the fact that critical or even hostile attitudes towards interracial unions began to shift in the mainstream around the 1990s, many millennials will still have memories of being referred to with offensive labels such as 'half-caste' and 'mongrel'.

Miri Song reiterates that the common supposition that mixed people are not quite whole is supported strongly by notions of blood quantum, comparing American usage of terms like 'mulatto' and 'octoroon' for African Americans to the aforementioned 'half-caste' for mixed Black and Asian people in the UK. Both Song, and Korean American writer and African American literature professor Julia Lee (Lee, J., 2023), have pointed to the manipulation of the 'one drop rule' in American history to enforce racial hierarchies. This was a legal principle of racial classification used in the USA in the 20th century. For example, it maintained that someone with one African ancestor ('one drop') was considered Black. The rule was not only used to suppress and control the civil rights of Black people in the USA, it was also used to forcibly displace and incarcerate at least 125,000 Japanese Americans during World War II, during a shameful and, at the time of writing, still very under-documented part of US history.

So it's not that surprising that, particularly for those of us born before the 2000s, there's a certain amount of scepticism and hesitation when it comes to the 'trendiness' of mixed ethnicity. There is a myth that mixed people are the cure for racism, and that their very existence is proof that we now live in a post-racial society. Besides the fact that many parents of mixed children can actively display and perpetuate problematic or downright racist behaviour, this myth also ignores the fact that racism is kept alive by deeply rooted systemic drivers, by institutions and practices deeply embedded into the fabric of our global societies.

MINORITY IN A MINORITY

'Being Viet Hoa means you've been an immigrant a few times,' explained Huong Black in the 2021 besea.n digital storytelling exhibition *Land, Sea & Stars* (Warner, 2021b). This project explored the stories of forced migrants impacted by the history of conflict in Vietnam, but as Huong deftly observed, previous generations of her family had already been uprooted from China to Vietnam before setting foot in the UK. Taking the Chinese diaspora as an example, simply assuming someone who identifies as Chinese has come directly from Mainland China would overlook the vast global journeys that its people have taken.

As well as Viet Hoa (which refers to Vietnamese nationals who have full or partial Han Chinese ancestry), there is also a sizable Malaysian Chinese population in the UK, bringing with them a varied set of influences, languages, perspectives and cuisines. Chinese Indonesians make up the fourth largest group of overseas Chinese after Thailand, Malaysia and the US. Indonesian artist Victoria Kosasie, spoke to us of the hardships faced by her family, including her Chinese father and East Javanese mother. The Chinese Indonesian population was oppressed through the outlawing of Chinese celebrations, speaking Chinese in public, studying in Chinese schools or even having Chinese names:

> You either had to assimilate, or suddenly declare yourself WNA (warga negara asing, or foreign citizen). Many families chose to assimilate, but continue practising their culture in secret – because how can you just suddenly go back to China, when you've never been there, and your family has been here for generations? (Kosasie, 2023)

Victoria describes how her grandparents chose to make an Indonesian sounding name to assimilate. Her father would insist at school that he was Dayak (a native Kalimantan tribe who also tend to have monolid eyes) rather than Chinese. She recalls that her father would constantly rub his eyes, an old habit from his youth that he felt would conceal his monolid eyes: 'Now, post-New Order era, my father chuckles as I embrace my Chinese roots by listening to Chinese music, watching Chinese dramas, and also as my sister populates magazines and Indonesian advertisements as a model and actor.'

Cross-cultural, often transnational, experiences allow for a greater sense of flexibility; out of necessity, we must make room for multitudes. In recent years, since the handover of Hong Kong from the UK back to the governance of China in 1997, many Hong Kong citizens have taken advantage of the British Nationals

Overseas (BNO) Visa to make a home in the UK and escape what they see as the return of restrictive social policies and human rights violations. While many migrants from Hong Kong who arrived in large numbers during the 70s and 80s may still identify as Chinese, the term 'Hong Konger' is gaining prominence (Lam, 2014), with a large spike noticed in Google search terms during the 2019 protests in response to the Hong Kong government's introduction of a bill to amend the Fugitive Offenders Ordinance with regard to extradition (Google Trends, 2025).

One interviewee, who wishes to remain anonymous, moved to the UK for school at age 15. They shared how their sense of identity has been impacted by moving to the UK later in life as a teen, admitting they had never experienced feelings of strong affiliation with nationhood:

> When people ask where I am from, I often will say, 'Hong Kong,' and when people ask, 'Are you Chinese?', my response will always be, 'Yes, I'm from Hong Kong.' While I was nurtured by elders with deep respect and understanding for Chinese culture, I feel it's important to acknowledge that an upbringing in Hong Kong would've been starkly different to that in Mainland China. (Anonymous, 2023)

While this person's reclamation of what it means to be from Hong Kong and having a sense of Chinese identity is strongly influenced by their time growing up in the country, for many who were born in the UK and who have rarely set foot in the countries of their forebears, this sense of connection happens via a kind of cultural osmosis. While society doesn't necessarily consider this a 'mixed' experience, it still uniquely places them at a cross-section of overlapping identities.

Amy, co-author of this book, was often not aware of the way different cultural influences growing up informed her ability to relate to a cross-section of people upon leaving home. Although her Chinese parents and older siblings are migrants from Vietnam, they settled for a short period in Hong Kong before settling in South London. In her words:

> In our four walls, it wasn't unusual to hear the incessant switching between Cantonese and Vietnamese languages. I'd eat homemade phở and bánh cuốn as often as black bean chicken. At the time, I didn't realise that this would enable me to bond with future friends with links to Hong Kong, China and/or Vietnam, who similarly felt isolated and craved an understanding of what it was like to grow up with such a cross-section of cultural sights, sounds and tastes.

This cross-section often makes simplified labels such as 'British Chinese,' 'Chinese,' and 'Viet Hoa' feel insufficient. With a Viet Hoa mother and Welsh father, Huong Black encourages her kids to use 'Welsh Viet Hoa' as a way to highlight the diverse locations from which their parents have journeyed. For Karlie Wu, an artist and co-founder of besea.n, introductory labels can change according to the audience: 'It depends on the context of who is asking. I can use Scottish Hong Kong Chinese, Glaswegian Hong Kong Chinese Diaspora or Hong Kong Hakka Diaspora' (K. Wu, personal communication, 25 February 2024). The Hakka, literally meaning 'guest families,' are not named after a geographical location and are commonly understood to be descendents of those who fled social unrest and invasions throughout Northern China for many centuries. They found sanctuary in Southern Chinese regions, becoming 'guests' of the local Cantonese populations, hence the name.

Muslim-Malay-Singaporean theatre maker, Faizal Abdullah, urges us to resist categorisation from singular labels, unveiling the rich complexities of his ancestry beneath the stated race on his legal documents:

> … even if the race stated on your identity card (IC) is listed as Malay, it is often not as 'neat' as that, you might have Bugis or Javanese or ancestry. I have Arab, Javanese, Peranakan-Indian and probably some other ancestry that I'm not aware of, but on the IC it's written simply as Malay. (N. K. Ramli & F. Abdullah, personal communication, 15 June 2023)

Curiosity should also extend towards understanding countries beyond their geography. Singaporean creative producer, Khai, comments on people not understanding where to put her, mostly assuming she is Filipino and works at the hospital based on her appearance. Or, since she is Malay, they assume she is a Mandarin-speaking Malaysian. This general lack of awareness is most evident when it comes to ignorance in the different naming conventions outside the West. Khai, whose full name is Nur Khairiyah Binte Ramli, recalled her experience of banking in the UK:

> [Getting] a bank card was really difficult because they weren't sure what Binte [a patronymic][1] is. It was a circus … because Binte is not my first name or last name or surname. People [also] started calling me Mrs. Ramli, which makes it sound like I am my father's wife. And even that is a stretch, because Muslim women don't take on the husband's name after marriage, we keep our name. So I'm always Ms. Nur Khairiyah Binte Ramli. (N. K. Ramli & F. Abdullah, personal communication, 15 June 2023)

Difference in cultural values means a majority of ESEA naming conventions begin with family name to reflect the prioritisation and respect for origins and ancestors in contrast to the individualistic

---

[1]A patronymic is a name derived from the given name of a father or ancestor, often used to signify familial lineage or descent.

nature of Western naming conventions in placing first name before family name. This collectivism-individualism dichotomy traces back to Eastern traditions like Taoism that tend to focus on concepts of unity, whereas Western philosophers emphasised freedom and independence. Table 5.1 contains a non-exhaustive list of naming conventions across Malaysia to illustrate just how varied the approaches are, even within a country, especially when race and religion intersect (Robson, 2017).

Table 5.1   Naming conventions in Malaysia

Group	Order	Example
Malaysian Malay	Given (double) name, patronym (Bin or Binti/Binte), SURNAME	Nur Aisya Binti RAZIF
Malaysian Chinese	SURNAME, generational name, given name (romanisation of Chinese characters)	KAM Kia Wei (from 甘家伟)
Malaysian Peranakan	SURNAME, generational name, given name	CHO Beng Huat
Malaysian Chinese Christian	Christian given name, Christian middle name, SURNAME, Chinese given name	Alexis Michelle LEE Yue Xi
Malaysian Indian	Given name, patronym (a/l or a/p)[2], SURNAME	Nagaratnam a/l SUPPIAH
Malaysian Indian Sikh	Given name, religious name (singh/kaur)	Karamjit Singh
Malaysian Orang Asli[3]	Given name, patronym, SURNAME	Sagong anak[4] TASI

[2]a/l and a/p are abbreviations of 'anak lelaki' and 'anak perempuan' respectively, which translates to 'son of' and 'daughter of'.

[3]Orang Asli translates to 'original people' in Malay, adopted by the Malaysian government in the 1960s to replace the English term 'Aborigines' as detailed in Endicott, 2016.

[4]'Anak' translates to 'child (of)' in Malay.

Some more naming conventions across ESEA cultures beyond Malaysia include those shown in Table 5.2.

Table 5.2 Naming conventions in other ESEA countries

Language	Order	Example
Singaporean Chinese	SURNAME, Chinese given name, English given name	TAN Mei Ling Emily
Singaporean Malay	Given name, patronym (Bin or Binte), SURNAME	Khairiyah Binte RAMLI
Japanese	SURNAME, given name	KONDO Marie
Korean	SURNAME, given name	KIM Jong-un
Thai	Given name, SURNAME	Thaksin SHINAWATRA
Indonesian	Given name/s, sometimes SURNAME	Prabowo Subianto DJOJOHADIKUSUMO
Burmese	Names should always be in their full form	Aung San Suu Kyi
Vietnamese	SURNAME, two-part given name	NGUYỄN Phú Trọng
Filipino	Given name, SURNAME	Rodrigo DUTERTE
Cambodian	SURNAME, given name	HUN Sen
Laotian	Given name, SURNAME	Bounnhang VORACHITH
Timorese	Given name/s, SURNAME	Francisco GUTERRES

Source: Asia Media Centre, 2019

For Amy and Mai-Anh (co-authors of this book), borders and migration have influenced their family names. When Amy's family arrived in the UK from Vietnam via the Hong Kong refugee camp, her mother was unsure whether to give her Chinese or Vietnamese name. Furthermore, in Chinese and Vietnamese cultures, women don't tend to take their husband's names. However, UK border control, unable to comprehend a married woman keeping her own name, marked down her Vietnamese first name and her husband's Chinese family name. For as long as Amy can remember, her mother has had a chip on her shoulder about this.

Mai-Anh's mother, also concerned about questions at border control, decided that every family member having her father's patronymic Scandinavian surname would be institutionally and practically easier. Because of this, her parents were insistent on their children having Vietnamese first names and refused to choose easier, 'Western' names for them, despite the fact that Vietnamese pronunciation trips up the average anglophone speaker. Mai-Anh has since changed her surname back to reflect her full heritage and, in recognition of the tradition of her mother's homeland, did not take her husband's surname on marrying.

Perhaps a starting point is to consider an alternative where legal documentations are not divided into first and last name, but preferred and legal name in full to accommodate the diversity of how we prefer to share our names. This includes any desire to bridge Western and native names, instead of resorting to convenience and conformity. We have demonstrated just some of the infinitesimal microcosms that can exist within 'single' ethnicities. Imagine how many further pockets of identities there must be among such a vast array of ESEA backgrounds and you'll think twice before insisting to know, 'Where are you from?'

PARENTING ACROSS CULTURES

How can we put ourselves in the shoes of the first generation migrants, who dived head first into a completely different life across the world? Adapting to life in the UK and raising families in a new country came without a manual and many parents did the best they could to put food on the table and give their children good prospects.

It was not an uncommon experience for many children of these first generation ESEA parents to experience adultification, becoming responsible for reading and translating their mail, sometimes even filing their tax returns (Hui, 2023). Conversely, parents themselves can sometimes be infantilised, particularly at times when a limited English ability fails to convey a complex background and bilingualism in other languages.

Keira Sussex, a parent and community activist, shared their experiences as a queer, mixed white and Chinese person who was brought up partly in Hong Kong and moved to the UK in childhood. Heritage plays a big part in raising their child, taking into account their Hong Konger, Singaporean, Welsh, Jewish, East London and Essex histories:

> I want her to understand where all of her comes from, her whole self. Some who have met her would say she looks white, but she personally identifies as mixed, because it's so complicated for her to go [through] all of this heritage. (K. Sussex, personal communication, 9 August 2023)

Lifting the veil on where exactly our roots lie is often fraught because language and generational barriers exacerbate the distance we feel from our past. How can we hope to pass on a sense of self if we aren't exactly sure of the places in which that self lies? The sense of cultural severance can sometimes wreak havoc on a parent's confidence in passing on a second language to their children as well. A so-called 'migrant tax' that comes in the form of extra language lessons, books and other resources adds an additional layer of financial burden, highlighting our earlier point about economic capital increasing access to 'ethnic options'. If time and money allow, holidays 'back home' can sometimes offer a sense of cultural grounding, but are limited to expensive school holiday periods. Instilling a sense of pride in ESEA identity can sometimes feel like an uphill battle, particularly if only one parent in the family has an ESEA background.

Navigating across cultures as parents can also look like mediating the grandparent–grandchild relationship, particularly if they live in separate countries. Wei Chieh Soon, a first generation Malaysian Chinese immigrant in the UK and parent of two, has been able to respectfully mitigate differences in opinion on parenting with his parents back in Malaysia by using physical distance to facilitate sensitive conversations. He has since aimed to bring openness and honesty to his relationship with his children, even if that means letting them know about conflict (W.-C. Soon, personal communication, July 2022).

If a mixed ESEA child is raised by two parents who do not have a lived experience of being mixed, there is now a great relief to have the benefit of those who amplify and share their knowledge so that children who may experience Othering can feel supported by their parents, guardians or carers. Writers such as Pragya Agarwal, author of *Wish We Knew What to Say* (Agarwal, 2020) and early-years antiracist trainer Liz Pemberton, who offers consultation and workshops, are all working to evolve conversations around intercultural parenting.

Keira's experience being analysed, questioned and judged as a mixed ESEA person growing up has allowed them to approach parenting with a sense of openness. Not feeling the pressure to have all the answers allows them to move forward: 'Being cut off from my Chinese heritage, having all of that identity taken away from me while still looking Chinese felt so damaging.' They expressed frustration that, to strangers, their physical appearance equates to knowledge of Chinese language and geography. This extends to how they want their child to be treated: 'I [don't] want my child to be an encyclopaedia of all of her heritage, but to have an understanding of all the things that came together to make her' (K. Sussex, personal communication, 15 June 2023).

Always searching but never truly finding the answers seems to be the crux of diasporic longing for many ESEA people, who want nothing more than to save their kids from questions surrounding belonging and struggles with identity. While some may have the benefit of extensive family trees and stories passed down through generations, it can often be the case that critical facts and information are obscured. Cross-cultural and mixed-ethnicity parents acutely understand that feeling of Otherness and feel better equipped to support their children while they navigate a complicated world. Huong Black offered this solution for her mixed children: 'I tell them to embrace the complexities' (Warner, 2021b).

CREATING A FUTURE OF EXPANSIVE, MULTITUDINAL BELONGING

What we hope we've achieved in this chapter is to illustrate the sheer complexity of mixed and multicultural ESEA experiences and shed some light on how further complex those experiences can be with global migration and/or displacement. Reconciliation with how these journeys have impacted us can take many forms. Therapy is sometimes necessary, and sharing our stories with those who have similar experiences, uncovering hidden truths or sometimes just acknowledging that we may never know. Healing comes through acceptance and the refusal to gatekeep what constitutes being ESEA.

The sense of global awareness and the way in which we can live contrasting ways of life is a testament to our potential for empathy with those who are excluded and marginalised, whether they are ESEA or not. It should make us feel proud of the many multitudes within our vibrant communities, and serve as a blueprint upon which we can build as we raise and nurture future generations. What we have seen is that imposter syndrome exists everywhere, yet we have the tools with which to construct welcoming spaces for exploration and curiosity – spaces that will hopefully offer our children the luxury of choice and agency as they settle comfortably into their own identities.

6

APPROPRIATION, EXPLOITATION AND WORK

INTERPRETING CULTURE

Culture and society are in a constant state of flux, deeply entwined with economic behaviours, being both influencers of and influenced by global economies. Cultural theorist Stuart Hall offered the explanation that different social groups create, share and struggle over interpretations of the world around them, primarily through the exchange of encoded and decoded signs and symbols within a given social context. Power dynamics influence how meanings are constructed and disseminated, especially in the media (Hall, 1973).

Culture, therefore, is always a translation, an interpretation.

This chapter looks at ways in which ESEA cultures have been interpreted in the media and in business, with a particular focus on marketing. However, it will also address root behaviours of colonialism, imperialism and extraction, and encourage reflection on how the exploitation of workers and resources in extractive practices and consumerism fund the economic and climate crisis we are currently facing.

First, we must make a distinction between colonialism and imperialism, as these terms are often used interchangeably. **Colonialism** is a form of imperialism that involves physical occupation and settlement. **Imperialism** can take many forms including economic control, military influence, or political pressure, without necessarily involving direct colonisation. Colonialism is often a tool of imperialism. Many imperial powers used colonialism as a strategy to expand their influence, exploit resources and control trade routes.

While colonialism largely declined in the 20th century, its legacy still shapes the world, and imperialism has continued in different forms, such as economic and cultural dominance (sometimes called neoimperialism or neocolonialism).

WHY COLONIALISM MATTERS

Extraction of resources and exploitation of land and people were common goals of imperial powers throughout history. The British led large deforestation projects and forced local populations of West Bengal into indentured labour on tea plantations in appalling conditions, which, despite some reform since independence, are still grim today (Roy, 2022). In Malaysia, the cultivation of rubber decimated natural landscapes and the manipulation of land policy resulted in discrimination against the Malay population (Light, 2020).

Meanwhile, the French colonial government made vast swathes of land in Southeast Asia (then French Indochina) available to private companies for rubber plantations, from which it stood to profit due to favourable tax and export tariffs. Survivors of the indentured labour regimes on French rubber plantations, such as Trần Tử Bình, described the conditions as 'hell on Earth'. Writing in his 1985 memoir, *The Red Earth*, Trần describes the poor living and working conditions, regular beatings, malaria and dysentery endemics and intolerable cruelty as 'life imprisonment without a jail' (Trần, 2014).

Today, the exploitation of land and resources is still ongoing in some places, such as in Occupied Palestine, where the forced erasure of Palestinian agricultural society by Israel is carried out through the restriction of land access and of resources like water, the destruction of agricultural means of production such as olive trees, and the planting of forests to obscure evidence of Palestinian agricultural existence (Golzar Anderson & Phan, 2022). When we come onto the subject of cultural appropriation later in this chapter, we must bear in mind the foundational principles of colonisation projects: the imperial power exerts power over the colonised and extracts resources for capital gain.

While some instances of colonial land and resource exploitation are still in practice today, in many countries with former colonies, the hangover of imperialist exploitation is visible via softer outputs, such as cultural and historical ones. Many museums, for example, have legacies rooted in imperialism, with collections made up of artefacts looted during colonialist expeditions and conflicts. Not only are priceless pieces of art and local history kept thousands of miles from their places of origin, some museums even display looted body parts (Chamberlain, 2021).

Numerous modern-day countries, like Sudan, Nigeria and Benin, have made calls for former colonial powers to return stolen artefacts, with varying degrees of success. Campaigns to decolonise museums have been met with both support and resistance from within the curatorial sphere, with some concerns that to decolonise is to decontextualise or rewrite history. This historical exploitation is further compounded today by some institutions' failure to engage meaningfully with impacted communities on new exhibitions, as we saw with the allegations of copyright infringement levied against the British Museum (see Chapter 4: Recognition without redistribution: Who truly benefits?).

Palestinian academic Edward Said's (1978) concept of Orientalism, a critique of the West's perception and portrayal of the East, or the peoples of the Orient[1], helps us frame our understanding of the ways in which certain cultures are interpreted popularly today. Orientalist outputs such as art and literature are often characterised by stereotypes, exoticisation and a tendency to view the Orient as mystic, homogenous and inferior.

Today, we regularly see Orientalism through marketing campaigns. In 2014, Air France launched a campaign called 'France is in the Air', featuring 12 visuals depicting iconic destinations and models dressed in what one blogger called 'cultural drag' to match (Jenn, 2014). The visual for Tokyo is particularly striking: a white model, decked in full Geisha costume and makeup, complete with hair sticks, clutching a piece of fruit. Air France customers arriving in Tokyo are not likely to see the streets of Japan's capital, one of the most populous cities in the world, lined with demure geishas and pomegranate trees.

Meanwhile, in 2021, restaurant chain The Ivy announced the opening of The Ivy Asia in London with a video advertisement depicting heavily Orientalised caricatures, dressed in a mashup of Chinese and Japanese costumes, bursting into

---

[1]North Africa and Asia, including Southwest Asia, more commonly known by the European imperialist term, the Middle East.

a civilised, classy establishment full of smartly dressed white diners (Layput, 2021). The characters speak gibberish, reminiscent of the unintelligible noises often used to poke fun at East Asian languages, and become quiet (and presumably civilised) once they sit down to eat in the restaurant, the implication being that The Ivy Asia brings class and sophistication to ESEA cultures, without the barbarism. After immediate backlash on social media, the video was removed, but not before it was captured and re-uploaded on YouTube.

The problem with these marketing campaigns is not only that they flatten and homogenise cultures, perpetuating harmful stereotypes, but they often do very little to serve the countries or communities from whose cultures they extract. Furthermore, the borrowed elements rarely go further than superficial aesthetics. We should ask ourselves: who profits from these aesthetics? Do they really celebrate and show appreciation for cultures? Or are our customs simply a prop to win clout and profit, unlikely to be funnelled back into the communities that have been appropriated?

APPRECIATION VS. APPROPRIATION

We've all seen it: the white guy with a tattoo that reads 'peace' (supposedly) in Japanese, the blonde girl in a sushi advert wearing a cheongsam-style minidress and chopsticks in her hair, festival-goers at Coachella wearing bindis. Over the last couple of decades, Western pop culture seems to have been telling us something: Asian stuff is *cool*[2]. But does this rise in

---

[2]We are using 'Asian' with intention here, in recognition of the fact that almost all cultures across the continent have been subject to exploitation, and exploitation by Western powers.

popularity in the West[3] of Asian art, fashion, music, food and traditional practices mean that Asian people are gaining more cultural and social capital? What material implications does this have for historically excluded communities?

Cultural appropriation is most commonly understood as the act of members of a majority group adopting cultural elements of a minority group in a way that is exploitative, disrespectful or reductive. One of the popular arguments against claims of appropriation is that, in a broad sense, all creative expression is derivative, and to be inspired by another culture or aesthetic is not to pay compliment to the original, but to follow a normal part of the creative process. Furthermore, the trend of ESEA cultures inspiring Western creators is not always directly a result of colonialism. For example, when Japan ended its isolationist foreign policy in the mid 19th century, Japonisme (the popularity and influence of Japanese culture in European art) was not solely extractive, with Japanese artists themselves shaping their creations in response to the growing attention they were receiving abroad. Influences spread beyond art, from Pierre Loti's *Madame Chrysanthème* (the precursor to opera *Madame Butterfly* and musical *Miss Saigon*, both widely criticised for Orientalism) to the Doig Pagoda, an East Asian-inspired architectural feature used in more than 50 distilleries across Scotland.

Fast forward to the digital age. Just look at the names of the software and products you may use every day: Asana,

---

[3]Some Asian pop culture like Japanese and Korean cinema, music and fashion are popular all over other parts of Asia, but since this book is about ESEA people in the UK, we're focusing on the power dynamics of the Global North and cultural exports.

Guru, Genie, Gong, Ninja. In case we were in any doubt, the tech world has reinforced all the trappings of Asian mysticism, but with added sprinkles of model minority efficiency. In the fashion industry, the use of Chinese and Japanese characters in clothing lines has been especially popular, with SuperDry being one of the most famous brands – despite being Japanese branded, it has no connections to Japan and the use of kanji is often random or grammatically incorrect. Meanwhile, the ESEA aesthetic has become such a common feature in branding that there is a ubiquitous and recognisable 'Asian font.'

Are Asian people finally getting recognition in the mainstream (not that it should be an aspiration), or is this just modern day Orientalism? The popularity of some concepts seems harmless enough, even positive, such as the preference for feng shui concepts in interior design, and sometimes it's amusing, like using culinary chopsticks in hair styling rather than hair sticks.

There are plenty of examples of cultural borrowing that cause personal harm, however. In 2021, activist and community organiser Kimi Jolly discovered a black and white photograph in a shop, of two Kelabit men with traditional elongated earlobes and jewellery, both topless. She was stunned to realise that one of the men was her grandfather. Further research led her to find that, along with a photo of her grandmother, the photo had been sold to a company that makes prints and photographs, and these were widely available to purchase as souvenirs from Borneo. Neither Kimi nor her mother are aware of any attempts to contact the family to ask permission to distribute or sell the photographs, and they do not receive any royalties.

Kimi has since discovered pictures of her grandmother on strangers' Instagram pages, with particular attention paid to her earlobes and traditional tattoos, and feels frustrated at the violation of privacy and lack of respect. 'It's as though they're pointing out a feature of a wild animal in a nature documentary,' she said. 'They don't know my family's history. It's nice to see the photos when people are respectful, but I hate the thought of [them] being used as postcards without people knowing who they are or their life stories' (K. Jolly, personal communication, 14 January 2021).

Southeast Asia is a popular travel destination for Western backpackers owing to its cheap cost of living, tropical climate and extremely varied landscapes. However, in the last couple of decades, it's also become something of a hedonistic playground, attracting travellers who want to check popular tourist activities off a list without paying much money, or much attention to the impact on the local population, their traditions and cultures (Hampton & Amran, 2016). With the rise of digital technology and our society's penchant for gratuitous, real-time posting on social media, Kimi's story is a reminder of the importance of slowing down, remembering who is on the other side of the camera, and evaluating the dynamics of power that accompany every social transaction.

CULTURE UPGRADED

It is one thing to borrow and profit, but it is another entirely to claim superiority. In recent years, we have seen the rise of various non-ESEA ventures claiming to improve or revamp traditional ESEA cultural practices in order to boost their popularity among non-ESEA audiences. In late 2020 and early 2021, during the grip of the COVID-19 pandemic and a wave of anti-ESEA hostility and violence, the founders of USA-based company The Mahjong Line came under fire for their attempts to 'refresh' the ancient Chinese game at an extortionate price tag, with tile designs they saw as more relatable to the average American, the implication being that the average American, to them, is white (Chen, 2021).

2021 also saw the dressing down of self-proclaimed 'Congee Queen' Karen Taylor, the white founder and CEO of the company Breakfast Cure, which boasted a modern, improved version of congee (a popular rice porridge eaten across Asia). Instant pot packets of congee, which serve four people, retail for $15 on the company's website with a $9 delivery fee, boasting flavours such as 'Apple Cinnamon' and 'Blueberry Bliss.' Traditional congee is usually savoury, although ingredients vary by region. The price tag adds insult to injury, since, in many countries, congee has humble roots. In places like Vietnam, for example, congee saw many people through times of economic hardship, since the high ratio of water in the dish means a small amount of rice can go a very long way. Both Taylor and the owners of The Mahjong Line were accused of whitewashing cultures and making them less 'foreign' in order to appeal to the average American consumer, facing widespread criticism and outrage from Asian communities and beyond (Mark, 2021). They quickly corrected the course, updating their branding and pledging money to various initiatives supporting Asian American communities.

While the founders of The Mahjong Line and Breakfast Cure did apologise, the perpetual cycle of offend, rinse and repeat can be exhausting for impacted communities. Furthermore, falling into the dangerous habit of causing offence and then simply making amends with your wallet is not a sustainable path for true understanding and appreciation. The appropriation of cultures for profit is not simply an annoyance for Asian diaspora in the West: in the face of decades of colonisation, systemic exclusion, alienation and isolation, many communities cling to their traditions and cultural practices as a routine means of resistance to oppression.

The list of cultural appropriation offenders in the culinary world is lengthy, such as Itsu, Pho, Hop and Tofoo, to name a few. Tofoo is a Yorkshire-based company that makes tofu that it claims to be 'a zillion times tastier', according to its website. The claims of superiority go even further: 'Tofoo's not the soggy, tasteless stuff you might be thinking of, and we're on a mission to show the world that tofu is really rather brilliant. Actually not all tofu, just ours.' In perhaps a more outrageous display of audacity, Pho, the popular white-owned Vietnamese food chain, made headlines in 2013 when the owners trademarked the word 'pho' and attempted to sue a family-owned Vietnamese restaurant for using the word in their name (Hyun, 2019). Family-run ESEA establishments, often an employment lifeline for immigrants, are the subject of increasing speculative raids under increasingly hostile and carceral approaches to immigration, no matter the government in power.

Such companies will claim to be bringing the joys of far-flung corners of the world to the West, but are their consumers really learning anything new about the cultures of origin? And are they learning it from the right people? It is noteworthy that these companies often benefit from the more positive experiences of the appropriated cultures such as Lunar New Year, but fail to show up in ways that support in any meaningful sense. One might ask the question 'How can I carry out my business or campaign in a way that is respectful to the cultures I am borrowing from?' but perhaps the very first question needs to be 'Do I need to be the one to carry out this business?'

Furthermore, we mustn't underestimate the ways in which extractive and appropriative processes have been weaponised to fuel the interests of capital at scale. A particularly egregious example is the way in which soy beans – a vital part of many Asian culinary practices – were introduced to Western economies in the 20th century in response to a need for oil-based production and livestock feed. Environmental historian Jessica J. Lee explains that the popular perception of tofu as bland, soggy and unpalatable among Western imaginations is a direct product of the marketing efforts of the American mills and companies that profited from the non-culinary production of soybeans. According to Lee, the likes of Henry Ford, who invested in soybean research to fuel production of plastics, car frames and other goods, would not have been able to protect their trade at such scale if a demand for tofu as culinary produce had been allowed to thrive (Lee, 2024).

It is, of course, completely possible to appreciate and enjoy things from different cultures in an intentional and respectful way, and we should be wary of the tendency to gatekeep. It is one of life's greatest joys and privileges to learn new things about other cultures, and, indeed, most Global Majority cultures around the world are proud to share their food, designs and music with the world. Cultural export is a way for countries to develop their soft power – a term coined by Joseph Nye in 1990 at the end of the Cold War, which describes a country's ability to co-opt rather than coerce, using cultural output and ideology to improve international relations and standing – as opposed to hard power methods like military force (Nye, 1990).

At the time of writing, there is perhaps no country doing this with more intention than South Korea, which is one of the only countries in the world with a dedicated goal to becoming the world's leading exporter of popular culture. The phenomenal growth of the Hallyu, or the Korean Wave, in the popularity of Korean food, music, cinema, television, art and fashion, contributed a whopping estimated $12.3bn boost to the Korean economy in 2019 (Roll, 2021). In the school of thought of Italian philosopher Antonio Gramsci, soft power is an important tool in maintaining capitalist hegemony and securing cultural and ideological dominance – so it's perhaps not surprising that the Hallyu has accompanied an explosion of Korea-focused consumer frenzy, making the cultural shift somewhat bittersweet.

How can we appreciate and celebrate diversity of creative endeavours and cultural exchange, *and* shift away from growth-focused models of consumerism fuelling the climate crisis and increased global inequality?

POWER DYNAMICS

While convenience and scarcity inevitably influence our choices, it's also true that small, immigrant-owned businesses are more likely to struggle. In a world with a growing number of socially conscious consumers, we must remember that paying respect to the cultures we enjoy is an important part of the principles of sustainability and ethics that underpin conscious consumerism.

Wherever possible, we can choose to support local, independent businesses, which are more likely to struggle in today's economy. And we can make choices – from the clothes we put on our back to the cultural practices we engrain in our daily lives – to honour the roots of those practices and model consumer behaviour that aligns with values of anti-exploitation. We should be mindful that 'appreciating', in any true sense, goes a lot deeper than correct names or certain ingredients. We should consider whether only the cultural origins of the creators matter – what of gender, what of class, what of religion?

There's a significant difference between purchasing locally-made crafts on holiday and displaying them proudly in your home to remind you of your trip, and dressing up in a kimono as your Halloween costume. While it may seem like a harmless gimmick, and one might wonder if the same offence would be invoked when dressing in, say, rodeo apparel, we should ask ourselves about the balance of power. Is there a double standard in the protection afforded to the borrower not normally enjoyed by the people of the lending or appropriated culture?

A striking example of this double standard – afforded mainly to white people – was starkly obvious during the rise of the foxeye trend in 2020. The makeup trend, which usually involved a winged eyeliner application, was exaggerated in photos by the model placing their hands on their face and pulling back the skin around the eyes, to elongate the makeup effect. While many saw a chic pose, most ESEA people saw a triggering imitation of the gesture used to poke fun or bully those with facial features common to those with ESEA or Central Asian heritage. What made the foxeye trend most painful was its timing: in 2020, the rise in instances of COVID racism isolated, harmed and intimidated many ESEA people outside of Asia during the pandemic.

When we compare the imbalance of power and the stakes on the table for each group when it comes to the appropriators and the appropriated, it's easy to see that the potential harmful consequences of appropriation usually outweigh the minor disadvantages of not doing it. If the choice not to wear someone's traditional cultural dress as a costume – designed, frankly, to entertain – means that you're only mildly inconvenienced, it doesn't seem like such a huge sacrifice to make.

MARKETING ORIENTALISM AND THE CLIMATE CRISIS

Above all, branding and marketing that rely on the superficial borrowing of cultural ideas, aesthetics and practices ultimately serve the wider function of feeding up a consumer culture. We are increasingly having to reckon with the consequences of a globalised consumer economy and its impact not only on our natural world but on the people who live in it. As such, we must cast our critical net wider and consider the harm caused to those outside the digital Western pop culture bubble in which trends like foxeye and Asian baiting are allowed to thrive.

The country that produces the most consumer goods, at the time of writing, is China (CEPR, 2024). While most people acknowledge that the climate crisis is disproportionately impacting people in the Global South, with those in countries that contribute the least to global emissions suffering the harshest environmental damage, China is often overlooked because it has the world's highest CO_2 emissions (Carbon Brief, 2023). It's seen as a country that pollutes, rather than as a country that suffers the impact. What we don't often think about is the sheer size of China's population and its government's focus on meeting energy targets: if we look at the emissions per capita,

China's are less than half that of the USA, which has a population more than four times lower than China. But what of the impact on the workers who produce the goods, and the inhabitants of the areas around the manufacturing sites?

China is often referred to as 'the world's factory', and the growth in consumer demand has had radical effects on its economy, environment and population. Putian City, in Fujian Province, in the southeast of the country, is often known as 'Shoe City', due to the sheer number of foreign-invested shoe factories established in the city since the mid 1980s. As of 2001, the number of foreign enterprises for shoes alone stood at 150, with a total annual output of over 100 million pairs, and the majority of the workers are women (Chan, 2015).

According to the 2001 study conducted by researcher Anita Chan (Chan, 2015), in addition to being filled with toxic hydrocarbon fumes, the factories were all found to use concentrated, volatile, adhesive glue containing benzene, toluene and xylene. Repeat use of adhesives with high concentrations of these three hydrocarbons is extremely dangerous for health, with severe implications for blood production and the body's central nervous system. Chan wrote:

> Long-term contact can cause minor symptoms of dizziness, loss of appetite, loss of memory and mental capacity, and major symptoms of anemia [sic] leading to death. Most effects are gradual and may eventually lead to cancer and birth defects.

According to Chan, the Sanitation and Anti-Epidemic Station conducts medical check-ups every year, but the constant turnover of employees restricts their ability to follow up. There's also a lack of education on the subject, and deliberate skewing of results. Although this evidence dates back over 20 years, the implications for cancer and birth defects are long lasting and would certainly continue to impact generations of Chinese children to come, and we can see the way that the intersection of socioeconomic status and gender produces a specific experience of oppression under rampant global consumerism.

When it comes to the climate emergency, the impact of manufacturing emissions on the Global South is just one way that the world's Most Affected People and Areas (MAPA) are impacted by a phenomenon called waste colonialism. Waste colonialism describes the pollution of poorer countries by richer ones, either through direct or indirect means, such as the manufacturing implications as mentioned, or the shipping of plastic, textile and electronic waste to landfill sites.

In 2018, when China refused to accept any more waste imports, rather than the exporter countries finding waste solutions on their home soil, exporters looked for somewhere else to dump their garbage. Consequently, several countries in Southeast Asia began to feel the burden, as Global North countries such as the United States, Canada, the United Kingdom and the collective EU carried out mass deals with local waste handling companies in Malaysia, Indonesia, Thailand, Vietnam and the Philippines (BenarNews, 2019).

Furthermore, climate justice conversations are often rife with sinophobia (see Chapter 1: Sinophobia), which obscures efforts to hold Global North countries to account in public forums. The reality is twofold: firstly, this discourse obscures the very real impact on some ESEA people both in diaspora and at home; and secondly, distracts us from Global North nations' appalling failure to meet climate targets in recent years, with the USA and UK among the top offenders.

While this topic deserves far more attention than the scope of this book will allow, we would do well to remember that behind every poorly-executed marketing campaign is a product or a service that, more often than not, adds another weight to the bar on the shoulders of those in the Global South. Understanding how Orientalism and cultural appropriation are manipulated for capital gain, and the wider socio environmental impact of that gain, helps us to make informed choices when we choose goods and services. We must ask: what is the impact of our purchasing decision? Who profits? Does my need outweigh the damage?

It is impossible to take a prescriptive approach to cultural borrowing – after all, that's how cultural traditions evolve. A prime example is the birth of diaspora Chinese cuisines in takeaway establishments across the globe, adapted to please local palates – which are often quite different to the cuisines cooked in ESEA localities with majority Chinese populations. Or perhaps the influence of French cuisine on the street food of Vietnam, with the nation's most famous sandwich, the bánh mì, reappropriating the bread of its former coloniser.

In both of these scenarios, the resulting shift was a response to an imbalance of power: in a way, adapting cuisines can be seen as a means of survival under colonialism or systems of injustice. And so we can understand the critical importance of context: hierarchies of power, depth of knowledge, community care, capital gain (whether social, economic or cultural). If you attend a friend's wedding in another country, in all likelihood, they'll probably be thrilled to see their guests embracing the local customs by wearing traditional dress. Somebody who gifts you a culturally important piece of clothing is likely to be pleased to see you wear it with pride, and in so doing, honour and spread awareness of their country's traditions.

A SOCIETY BUILT ON EXPLOITATION

An uncomfortable truth we must all come to face is the extent to which the consumer-driven society we live in and depend on is built by and on the labour of undocumented workers. Their long hours, dangerous work and low pay keep our super-markets stocked with fresh produce, our construction sites running at all hours and our restaurants thriving.

Journalist Hsiao-Hung Pai (Pai, 2008), who has spent years reporting on undocumented Chinese workers undercover, argues that our society and economy is completely reliant on and intrinsically bound up with a hidden army of labour, bolstered in recent years by the demands of global capitalism and show-ing no signs of subsiding. This reliance is so strong the British government should have no choice but to focus its energy on creating a workplace society that upholds the rights and dignity of all workers, regardless of immigration status. Unfortunately, according to Pai, we have only seen immigration policy changes that drive undocumented workers deeper underground – often into the world of gangmasters and dangerous, organised crime.

As for how migrants are shaped in our collective conscious-ness, the effect of institutionally hostile attitudes towards migrants should not be underestimated, especially when it comes to shaping a collective sense of responsibility and moral-ity. We have seen Britain's institutional and media rhetoric around refugees and irregular migrants become increasingly fascistic and devoid of humanity in recent years. These narra-tives have the power to influence our ideas of what is right and wrong, what and whom we choose to care about, whose lives we prioritise, and whose lives carry less value. Increasingly, there are tragic accounts of migrant deaths as a result of violent border policies, each case treated with varying degrees of disdain by politicians and public figures.

In 2004, (then sitting) Conservative MP Ann Winterton joked about sharks going for a meal in Morecambe Bay, a grotesque dismissal of the deaths of 21 Chinese cockle pickers earlier that year, who drowned in the rising night tide (Wintour, 2004). In another disgraceful display of abject disrespect, Jeremy Clarkson, in expressing his disdain for synchronised swimming in early 2012, described it as 'Chinese women in hats, upside down, in a bit of water ... You can see that sort of thing on Morecambe Beach. For free' (Lovell, 2014). But the issues surrounding their deaths run deeper than offensive comments made by some individuals.

In an interview for *The Guardian*, Jabez Lam of Hackney Chinese Community Services (now ESEA Community Centre) spoke of the rage felt by the Chinese community following the Morecambe Bay tragedy, with the case of the 58 Chinese migrants killed by suffocation in the back of a lorry in Dover just four years prior still fresh in their minds. Lam said that the government's own immigration policy was to blame: 'They are denying people the right to decent living conditions, and the right to work and housing. These people have to put themselves at risk to make a living' (Ward et al., 2004).

2019 saw the case of the Essex 39, in which 39 Vietnamese nationals, originally mis-identified as Chinese, were killed by asphyxiation due to the negligence of the traffickers transporting them in the back of an airtight lorry, in a desperate attempt to reach the UK. A 2021 study looking at European newspapers' first response to the deaths found that regardless of the newspaper's political alignment, 'linguistic choices tend to dramatise what happened, criminalise victims, and even presume the driver's innocence, with the international criminal network he is presupposed to be part of remaining only speculated on' (Gregoriou et al., 2021).

Kay Stephens, co-founder of daikton* zine and organiser for the solidarity group Remember & Resist, summarised the root of the problem in a 2019 article for the Institute of Race Relations:

> The mainstream's response – calling for harsher borders, criminal justice for 'greedy and unscrupulous' traffickers and safe passage for 'genuine' refugees – fails to interrogate the global conditions that lead people to risk dangerous travel, and the deadly effects of border controls on all migrants. (Stephens, 2019)

WORKERS' RIGHTS IN THE AGE OF THE HOSTILE ENVIRONMENT

While many came to the UK to work in hospitality, professional and educational services, a significant proportion of ESEA migrants to the UK are involved in health and social care, and domestic work (The Migration Observatory, 2024). Many workers have experienced the injustice of providing a critical and needed service and being rewarded with inequitable working conditions and discrimination. ESEA migration and community organising has always been, and will remain, deeply tied to workers rights.

Probably the earliest example of this migratory phenomenon is seen among the travelling ayahs or amahs; women employed by the British in former Asian colonies as nursemaids, nannies and ladies' maids. Often, these women would accompany a family to care for the children on the annual sea journeys to Britain, but since most wealthy families would have retained permanent staff at home, they were then expected to wait until needed for the return journey.

These women came from all over Asia: India, Ceylon, China, Hong Kong, Burma, Malaya and Java. Sadly, many families did not honour their commitments, and left the ayahs without pay or a return ticket home, leaving many of these women destitute, in search of other employment, or forced to beg for funds for a return journey. The Ayahs' Home in Hackney, a boarding house for those stranded or waiting for their return journey, was commemorated with a blue English heritage plaque in 2022 (English Heritage, n.d.). However, while boarding houses such as this one provided a space for the ayahs, they were run as commercial businesses rather than as philanthropic initiatives, and often provided an extension of a racialised, colonial space in which the women were Othered and patronised by the superintendent and his family (Robinson, 2018).

The treatment of the ayahs would not be the last time that migrant care workers would receive poor treatment for their service in Britain. Fast forward to 2020. The Philippines is the world's largest exporter of nurses, with a significant number employed within the NHS (around 40,000, making them the second highest ethnic minority group within the health service) (Lorenzo et al., 2007). The NHS is one of the most diverse employers in the country, thanks in large part to its international recruitment programme, which seeks out healthcare professionals abroad – notably in countries where English is the primary language of instruction for medical education, such as the Philippines, Malaysia, India and Nigeria. It is critical to understand that the large-scale migration of Filipino care workers stems from US imperial policy, which established an English-language education and nursing system designed to supply labour to American institutions – laying the foundation for the Philippines' modern labour export policies and creating a workforce to serve richer nations rather than its own (Choy, 2003).

During the early months of the COVID-19 pandemic, prior to vaccine development, it was widely reported that racialised healthcare workers across the US and UK were dying in high numbers, with suspected institutional racism a contributing factor. By May, approximately 22% of all NHS staff deaths were Filipinos. In addition, Filipino nurses faced open COVID racism from patients who did not want to be treated by ESEA staff (Mercer, 2021).

In an interview in 2020, clinical NHS research nurse Lloyd Nunag, who came to the UK from the Philippines in 2017, spoke of the disproportionate number of deaths among Filipino nurses (Nunag, 2020). He believes that a combination of structural racism and negligence lead to the high death toll, pointing to ill-fitting PPE as one example. Nunag also believes that some Filipinos and other migrant health workers found it difficult to refuse when asked by their managers to work in COVID-19 red zones without proper PPE. Like most migrants whose status depends on their work permit, they were afraid to risk doing anything that might compromise their ability to work and send money home.

The fear over conditional migration status – along with a potential for racial bias to assume Asian workers will keep their heads down and get the dirty work done – is a theory that many Filipino healthcare workers in organising spaces have put forward. In May 2020, the Filipino Nurses Association submitted written evidence to Parliament with a number of concerns around the striking number of deaths and infections, lack of adequate or properly fitting PPE, bereavement and family support, and, crucially, an investigation into why Filipino nurses were always assigned to COVID wards, and why even those workers who were high risk and should have been shielded were assigned to the front lines. A staggering 80% of the Filipino staff who died were in this category, at the time of the submission (Parliament, 2020).

While some action was taken by the NHS to address systemic inequality both in an acute response to the crisis (the establishment of a helpline for Filipino healthcare workers impacted by COVID-19) and longer term (the appointment of Oliver Soriano, the NHS' first ever Filipino Chief Nursing Officer in 2023), the bar is painfully low. It is no wonder that, when the state or other employment institutions fail to provide adequate protection, impacted communities turn to each other and create the support and advocacy they need.

Alongside organisations like the Filipino Nurses Association (FNA), new groups like Filipinos in Care, the first international nursing and midwifery association solely dedicated to Filipinos in the care sector, have sprung up in recent years to meet community needs. The FNA have since interacted with nursing unions to provide support to protect workers over establishment interests, and further reinforce the framework of workers' organising across different ethnic groups within the health and care sector.

Another group providing community lifelines is the grassroots, migrant-run organisation, The Voice of Domestic Workers (VODW) (see Chapter 4: The Unstoppables). Under the rallying mantra of 'domestic work is work', the organisation's principle goal is to support and educate around migrant domestic workers in precarious or vulnerable positions in the UK, such as those with no recourse to public funds or those trapped in exploitative or abusive employment. It does this through its core working groups dedicated to specific issues, such as trade union activity, education and welfare. According to the organisation, around 23,000 domestic workers are brought to the UK by their employers on private domestic worker visas.

It is through the treatment of migrant domestic workers that we see a painful consequence of the UK's Hostile Environment policy. It was introduced in 2012 by then Home Secretary Theresa May, to make life difficult for undocumented migrants in the UK. The goal was to deny access to essential services like the NHS and state support, make it illegal for them to work or rent property, and require doctors, landlords, police and teachers to check immigration status. The policy led to increased scrutiny of people who 'look or sound foreign' and allowed the Home Office to access personal data from public services, affecting even victims reporting crimes or seeking medical care (JCWI, 2024).

The Hostile Environment has had monumental negative impacts on the welfare of migrants in Britain, both documented and undocumented (despite May's phrasing of the policy, which was meant to target 'illegal' immigrants). It deters people from seeking medical care or help via other public services, and traps them in untenable and dangerous working and living situations that violate their human rights. Such is the case for many domestic workers, legally employed in the UK, whose visa criteria were changed under the Hostile Environment to remove a worker's right to change employers under the same visa.

VODW says that the effects of this policy change have left workers less visible, more vulnerable to abuse and 'unable to access employment law or even to withdraw their labour and so challenge abuse' (The Voice of Domestic Workers, n.d.). Workers are often poorly paid, overworked, physically or sometimes sexually abused, subject to racial bullying and deprived of dignified working conditions. In some cases, VODW has reported workers having to eat the leftovers of the families they work for. The organisation argues that the removal of these rights is not addressed by either the 2015 Modern Slavery Act or the 2016 Immigration Act, and does not reduce migration to the UK, but merely funnels it into irregular, underground networks that are unregulated and rife with exploitation.

The kind of visa precarity that traps migrants into high risk exploitative jobs is also something we can see in the reliance on migrant fruit picking labour under the Seasonal Worker visa. After Brexit and Russia's invasion of Ukraine, the UK has had to look further afield for farm workers, most significantly in Indonesia. Indonesian agricultural workers in the UK were at huge risk for debt bondage after British recruitment agencies relied on poorly vetted local agencies in Bali, who in turn used independent brokers to supply workers. These brokers then charged exorbitant fees to the workers they supplied (The Guardian, 2024b). To make the problem worse, the workers were employed on zero hours contracts, so they did not have any guarantee of having enough work to repay their debts.

Although recruitment in Indonesia has now been suspended, if the UK is going to continue to focus on international recruitment to meet its labour needs and obligations under the Modern Slavery Act, action is required from the government, the farms involved and the retailers (supermarkets like Tesco, Waitrose and Morrisons have all been implicated). It is critical to fill the monitoring and evaluation 'black hole' in supply chain management to secure safer routes and conditions for seasonal workers (Dugan, 2022).

Francesca also described the double-edged sword of being a migrant worker in a public institution like the NHS or a University, which require health workers to check immigration status or university staff to report attendance to the Home Office. In this way, people become 'both a migrant and subject to immigration enforcement, but also an agent of the state' (F. Humi, personal communication, 7 August 2023).

If not for the tireless campaigning and networks of organisations, issues like these might be fully obscured by lack of institutional due diligence, lack of media attention, lack of public awareness and general misperceptions around migrant workers. One can imagine that, in the case of ESEA migrant domestic workers, the intersection of gender (for domestic work is largely carried out by women) and race (the model minority myth again) could compound to further marginalise a group of already very vulnerable people who aren't often considered in public discourse on migration.

Art therapist and Anglo Scottish Filipino, Sarah Reid, spoke to us about her work with an ESEA migrant-centred charity:

> The word 'resilience' is very problematic. A lot of Filipino domestic workers are seen as being very resilient, but I think that sometimes just means 'suffering' and putting up with it, and it's a hard choice because they need to pay to support their families in the Philippines. (S. Reid, personal communication, 9 August 2023)

If migrant lived experiences are constantly couched in the language of hard work and stoicism, then it becomes harder to critique an immigration system that values some lives more than others. Ultimately, a lot of immigration systems in the

Global North rely on an ableist, capitalist mechanism that evaluates and values people only by their capacity to contribute positively to our 'society', which is really code for 'economy'. Through their daily actions, the leaders of VODW are dispelling the model minority myth and making it the norm for migrants to be loud, unapologetic and unyielding in their advocacy from the lower ranks of the social hierarchy up.

We must move away from the idea that migrants must be separate from activism due to a need to comply with their immigration conditions, and from human rights and solidarity movements that are tied exclusively to demands from the state. Instead, we must close ranks around the most vulnerable activists and use community power to keep each other safe. Anti-raids solidarity groups are a good example of alternative community protection, with increasing cases of blocked deportation flights and immigration raids harnessing power in numbers.

We have seen how shifts in policy radically impact the lives of those most marginalised in society, in stark contrast to the little impact on the privileged (here, we include not just the wealthiest 1%, but the comfortable middle classes). In 2024, the UK saw an end to a Conservative government that imposed fourteen years of austerity and increasingly hostile immigration policies[4]. What future governments will have to reckon with is a legacy of immigration laws that have seen increased raids that target commercial sites (based often

---

[4]It is important to say that, at the time of writing, it is not yet clear whether the new Labour government will scrap the austerity measures and hostile environment policies imposed by its predecessor, so change is not a given. Furthermore, the pre-2010 Labour is also responsible for its fair share of punitive immigration measures, which are often forgotten in the shadow of the Hostile Environment.

on racial profiling) and tight restrictions on safe migration routes, as well as an increasingly disillusioned and disenfranchised white working class population whose own struggles have been ignored, priming the field for right-wing populism. The Labour party has shown little sign of implementing serious changes, therefore reinforcing the need for acts of solidarity that endure beyond the transience of ineffective governments. With increasing mistrust of state mechanisms at the heart of many marginalised groups, at a community level, it is critical to focus on knowledge and resource exchange, mutual aid, collective care, and on what groups like VODW can teach other grassroots organisations about intersectionality and a truly inclusive politics of solidarity.

SPACE FOR PLURALITY

While we have encouraged the questions 'who profits?' and 'who's at risk?' when confronted with potential cultural appropriation or even outright exploitation, we might also remain vigilant of the pitfalls awaiting ESEA people themselves. Efforts to gather and organise are often centred around points of commonality, which can certainly reinforce community bonding, but there is a risk that such efforts can perpetuate certain stereotypes and contribute to a compression of different ESEA identities. Furthermore, there is a risk of homogenising ESEA cultures that often occurs couched in the language of unity and celebration. A clear example of this is seen in attempts to bring celebrations like Lunar New Year (or more accurately, Lunisolar New Year) or Chinese New Year into mainstream consciousness.

Firstly, many ESEA countries, notably in Southeast Asia, tend to celebrate solar rather than lunisolar holidays, which are very rarely acknowledged in any meaningful way in public spaces outside of Southeast Asia. Secondly, Lunisolar New Year campaigns and celebrations have a tendency to overlook the sheer diversity of the populations that celebrate the holiday. Different ESEA cultures interpret the zodiac in varying ways, serve different meals and honour different traditions. Not even all Chinese ethnic groups or diaspora celebrate in the same way, from the preparation of dumplings very popular in northern Mainland China and Chinese American communities, to the tossing of *yee sang* ('prosperity salad') across Malaysia and Singapore. Every family or household may also have its own traditions. It is also not uncommon that celebrations of different ESEA cultures in wider mainstream popular culture cater primarily to non-ESEA audiences, which is not necessarily a problem in itself, but one does wonder at the missed opportunity for meaningful cultural exchange in capacities that intentionally lean away from eurocentricity.

In conclusion, we should consider what's at stake in any attempt to bring ESEA communities and cultures into mainstream public consciousness and, indeed, if that should be a common goal. We need space for multiple approaches to community celebration and community strengthening: while some groups and individuals may choose to focus on widespread education and cultural diffusion, this should always be balanced out by the provision of dedicated spaces and resources to ensure the continued wellbeing, aspirations and longevity of ESEA people in communities, with collective care at the heart of each engagement.

7

QUEERING THE SCRIPT

REJECTING MAINSTREAM FEMINIST APPROACHES

Numerous Asian feminists have challenged the limitations of white feminism by highlighting how race, colonialism and imperialism shape gendered oppression. Chandra Talpade Mohanty has critiqued Western feminism for homogenising 'Third World women' and ignoring their diverse struggles (Mohanty, 1984). Grace Lee Boggs connected feminism to anti-capitalist and anti-imperialist movements, while Cathy Park Hong has critiqued the erasure, racial triangulation and the failure of the model minority myth in addressing the psychological and cultural alienation experienced by Asian American women, as well as the historical legacies of exclusion, Orientalism and labour exploitation (Park Hong, 2020). Their work provides us with the premise that feminism cannot be a 'one size fits all' movement – it must be rooted in the specific histories and struggles of marginalised communities.

A rejection of white feminism, which centres the experiences of white women while ignoring the ways race, class and other structures shape oppression, is essential. In order to understand how racialised misogyny is levied against ESEA women, and the performance of masculinity expected from ESEA men, we must ground our understanding in the tradition of Black feminist writers who have long argued that true liberation requires an intersectional approach – one that acknowledges how racism and patriarchy are intertwined, and that focuses on system dismantling and liberation for all, rather than on 'inclusion' within a flawed space. As Audre Lorde reminds us, 'the master's tools will never dismantle the master's house' (Lorde, 1984); feminism that fails to address white supremacy will never bring real justice for all gender-marginalised people.

Efforts to challenge harmful representations of ESEA people without the considerations above often risk reinforcing the very gender norms they seek to dismantle. Too often, responses to misrepresentation focus on replacing negative portrayals with more 'positive' ones, rather than questioning the racialised and gendered frameworks that produce these stereotypes in the first place. This risks reaffirming white supremacist ideals of respectability, desirability and gender performance, rather than challenging their foundations. Instead, we must ask: what does gender mean under these systems, and how might we refuse these imposed definitions? In this chapter, we dissect experiences at the intersections of gender and ESEA identity, and explore a queering framework – not to reinforce essentialist notions of identity and gender ideals, but to unsettle and disrupt the normative assumptions that uphold racialised patriarchy.

WHEN RACE AND GENDER COLLIDES WITH ORIENTALISM

We look again to Edward Said's work on Orientalism (see Chapter 6: Why colonialism matters), in which he stated that Western representations of Eastern men often depict them as either weak, effeminate and passive or, conversely, as hypersexual, violent and dangerous. This binary serves colonial interests by portraying the East as either in need of Western dominance or as a threat to be controlled. If 'Oriental' men are feminised, decadent, over-sexualised, or incapable of self-governance, then Western men, by contrast, are rational, disciplined, virile and decent.

Orientalist discourse has, over time, fed us various stereotypes that we will largely be familiar with: the weak intellectual, servile worker, asexual or hypersexual (depending on the context), despotic leader, martial arts hero, or ruthless businessman. For women, their bodies have been commodities and cannon fodder in the quest to colonise the world for centuries. A colonial tradition that treats Asian women's bodies as a metaphor for the colony itself creates, therefore, a site of colonial violence; it's a foreign, exotic land that must be penetrated, tamed and civilised. By now, the racialised misogyny that produces stereotypes for ESEA women is well known: Lotus Flower/China Doll, mail order bride, sex worker with no agency, heartless Dragon Lady.

These constructions have been shaped by Western colonialism, racial anxieties and economic shifts, reinforcing the West's sense of superiority. They have also heavily informed trends, fetish and sexual preference for ESEA women in particular, from pornography to dating apps. We wonder how many young, impressionable and insecure ESEA women have entered the minefield of the dating world, only to be met with suggestions that their romantic partners suffer simply from a case of 'Yellow Fever' (disregarding their individual appeal based on unique attractiveness or, dare we say it, personality). Or, how many have heard that familiar, taunting phrase: 'Me love you long time!' The cultural legacy of this scene from *Full Metal Jacket*, which reinforces Otherness and the association between Southeast Asian women and sex work, has outstripped the film itself.

The depersonalising, homogenising nature of racial fetish treats racialised individuals as replaceable and interchangeable. In the context of romantic relationships, this can bear a huge psychological burden on those targeted by fetish: it is exactly in the context of intimate relationships that it actually matters to be recognised as an individual with unique and irreplaceable qualities.

ON IMPERIALISM, SAVIOURISM AND THE ASIAN WOMAN IN DISTRESS

There is a long history connecting masculine, militarised narratives, Western exceptionalism and exoticised, feminine tropes in East and Southeast Asia, found in literature, film and theatre as early as the late 19th century (see: *Madame Chrysanthème, Madame Butterfly, Miss Saigon*). Some of the worst atrocities of colonialist and imperialist practices are obscured, forgotten or even forgiven – even today – in the face of the benefits of modern civilisation and military power brought to colonised sites through conquest. There's the common refrain that colonialism brought railways and infrastructure to underdeveloped lands, or that the political intervention and presence of Western military troops, as is the case with the Philippines, is protecting the people from a greater evil.

Rudyard Kipling's *The White Man's Burden* (1899) offers us a neat summary of the civilising colonial mission. Kipling's poem urges the then US president William McKinley to invade the Philippines, going on to explain the 'burden' of the white man to educate and civilise the 'sullen peoples'. The poem acts as a call to arms; a rallying, vocational quest. The expression 'the white man's burden' caught on and was used in product marketing and government documentation of the time, acting as a precursor to the more contemporary concept of 'white saviourism', which has found popularity in the language of identity politics today (sadly, often in watered-down ways that only serve to distract from the very real issue of Western exceptionalism).

In modern history, US military intervention in the Asia Pacific has echoed European imperialism, justifying itself as on a mission to 'save' locals while extracting resources and power. This has reinforced Orientalist tropes of racialised misogyny, reshaping the image of the Asian woman in distress. In East and Southeast Asia, she is hypersexualised and linked to sex work, while in West and South Asia, she is cast as a victim of patriarchal oppression – contrasted with the West's supposedly more 'civilised' patriarchy.

A normalisation of the militarisation of everyday life, where violence and control are seen as natural responses to conflict, reinforces patriarchal ideas of power, particularly when it comes to the site of that power. American presence in Southeast Asia during what was then known as the Second Indochina War (now, the Vietnam War) has shaped our global consciousness in a way that has endured long after the departure of the last troops – as a profoundly *American* experience in Vietnam. We call it the Vietnam War, although Vietnam has seen multiple wars in its existence (interestingly, in Vietnam, it's more commonly referred to as the American War).

The focus in films, tv and photography of and about the period is on American experiences, American lives, American attitudes and US military saviourism. The women left behind – think of the character of Kim in *Miss Saigon* – are not part of the main narrative. If we reframe that conflict to value the voices, narratives and experiences of Vietnamese people, which are few and far between, especially in the English language, we are left with so much possibility for new perspectives that disrupt the Western focus.

So how does this colonial shaping of social consciousness, particularly when it comes to ESEA women's bodies as hypersexualised and submissive, play out in the world of sex, dating and relationships?

A 2018 study found that pornography videos of Asian or Latina women were more likely to involve aggression than those featuring white or Black women and that violence towards Asian women in videos was most likely to be non-consensual (Shor & Golriz, 2018). Meanwhile, a 2021 report in the US found that Asian women made up 61.8% of the victims of reported anti-Asian hate crimes (Stop AAPI Hate, 2022), and writers like Giboom Park have suggested that sexual violence against Asian women in the US is extremely prevalent but under-reported compared to other ethnic groups, due to a combination of stigma, pervasive model minority ideology and collective shame (Park, 2020).

The epidemic of violence against ESEA women was brought into sharp relief in March 2021. On the afternoon of 16 March, a man named Robert Aaron Long murdered eight people at three spas and massage studios in Atlanta, Georgia, with a 9mm SAR semi-automatic pistol. Six of the victims were ESEA women employees – specifically, Chinese and Korean.

The gunman insisted that his actions were not racially motivated, but that he had a sex addiction, and had targeted the spas because he wanted to eliminate temptation. Captain Jay Baker of the Cherokee County Sheriff's office claimed at a press conference that the shooter had had 'a really bad day' (Mishra, 2021). Enforcement agencies were widely criticised, by both the public and by US representatives, for failing to treat the incident as a suspected hate crime, both on the grounds of race and gender.

While it's true that no clear racial motivation was given by the shooter, a failure to even interrogate the relationship between race, gender and class status overlooks the vulnerability of workers at this cross-section of society. A culture that roots power in violence and control is a dangerous thing, especially when bolstered by historical colonialism and a media that shapes pervasive racial stereotypes. It is rooted in the same white supremacist thinking that allows supposed law enforcement agencies to extend sympathy for a shooter's 'bad day' and 'sex addiction' and prioritise his humanity over the lives of his victims. While the shootings took place across the Atlantic, the grief rippled through ESEA communities in the UK. Vigils were held, and community groups documented how many ESEA women were retraumatised by the incident – not because it was unfamiliar, but because it echoed patterns they knew intimately.

ON DESIRABILITY, FETISHISATION AND SELF IDENTITY

One thing that's clear from the tropes and misconceptions we have explored is their profound impact on self-identity. The hypersexualisation of ESEA women, reinforced by colonial histories and media representations, creates a constant tension between visibility and vulnerability. How can one feel a sense of security in their body when society has already defined it for them – when their existence is filtered through stereotypes of submission, exoticism, or servitude? This sense of imposed identity does not only affect women; racialised gender expectations extend to ESEA men as well, shaping how they are perceived and how they experience their own masculinity.

Desirability, when examined through a racial lens, is also complex for ESEA men, with the relentless teetering between the extreme stereotypes of desexualisation and fetishisation, which can take a heavy mental toll. The same white supremacist logic that hypersexualises ESEA women often erases the sexual agency of ESEA men, casting them as either asexual and emasculated or as hypermasculine threats. These conflicting narratives have a deep psychological toll, creating internalised struggles with self-worth, attraction and belonging. The pressure to conform to white-dominant standards of desirability – or to reject them outright – becomes yet another battleground for identity.

This battleground is increasingly shaped by digital culture and the commodification of ESEA masculinity. We witness the growing fetishisation of ESEA men following the rising trend of Asian himbos and the Hallyu phenomenon. What is typically and narrowly conceptualised into the image of a thoughtful, chiselled oppa is amplified through social media virality, further entrenching one-dimensional portrayals of desirability. Content creators like YouTubers Jin & Hattie have capitalised on the Asianness of their romantic relationships by perpetuating hashtags like #AMWF (Asian Male White Female) whereas vlogger PrinceCheech strategically titled their most-watched vlog 'I finally met my KOREAN MILITARY BOYFRIEND *emotional*' to tap into existing fantasies of Asian masculinity (kimchiwangmandu, 2023). Yet, even as certain portrayals gain popularity, lingering stereotypes of emasculation and undesirability continue to overshadow moments of 'Asian excellence.'

For ESEA men, the struggle with self esteem and desirability is compounded when their reality doesn't match the images and expectations imposed on them. ESEA men who struggle with dating risk falling into a trap of shame and self blame at their lack of desirability, specifically blaming their race out of frustration. However, one study on online dating showed that Asian men actually face no more racial discrimination than their Black and Hispanic counterparts, despite the widespread belief that Asian men are the most penalised group of all men on the dating market (Curington et al., 2021).

Shame in this context can also bleed into internalised racism amongst Asian women as well as gay Asian men, expressing their own prejudice against dating Asian men (De Pacina, 2023). Research also indicates that Asian American women are more likely to prefer dating white men over Asian American men, as white partners are often perceived as a means of accessing 'proxy privilege' (Le & Ahn, 2024). Hostility experienced in this way has sometimes been cited as a cause of the rise of the 'ricecel' phenomenon. This is an iteration of 'incels' (involuntary celibates) where Asian men feel unable to form sexual or romantic relationships, throwing hatred and anger towards women, in particular Asian women, for their rejection of them.

In the most extreme circumstance, we witnessed the Isla Vista rampage that left 7 dead and 13 wounded. The perpetrator Elliot Rodger was angry about the fact that he had Asian heritage, which he perceived as an impediment to losing his virginity. Race was cited by psychologists as a key motivating factor for murder (Guillermo, 2014). Based on his 137 page manifesto leading up to his 'Day of Retribution', Rodger's struggle to integrate his Chinese and white heritage was evidenced as early as nine years old, when he asked his parents for permission to bleach his hair blonde. His preference for blondeness as a means of reaching for whiteness as wholeness is reflected in his views on white women and racial dating hierarchy, which stated clear beliefs in racial hierarchy and the desirability of Eurasians over 'full blooded Asian[s]' (Louie, 2014).

Rodger's crimes require much more consideration of class, colourism, toxic masculinity and misogyny than we have scope for, but we need to acknowledge that they stem from ethnic self-hatred, its intersection with gender, and a desire to feel accepted in society. The connection between incel culture and Orientalist eroticisation is a concerning one. Writing for VICE, Alia Marsha found posters on the Reddit forum for incels, a thread littered with misogyny and deep-rooted racism, claiming that the salvation to their rejection by women in the West lay in Southeast Asian women, particularly in Thailand.

In the 70s, Thailand was a common recreation site for American soldiers on leave during the Vietnam War. Today, it's still perceived as a playground for 'sexpats' and sex tourism. Marsha described an 'intersecting venn diagram of men's rights activists, incels, pick-up artists, and alt-right white nationalists that seem to have more than a thing for Asian women' (Marsha, 2018). The dangerous overlap between politics, sexual desire and racial hierarchy can have serious consequences with regards to interpersonal violence, as we've suggested, but is also linked to booming sex industries in Southeast Asia and, with it, the potential for exploitation. Meanwhile, a UN-backed report in 2016 linked the rising levels of sexual exploitation of children in travel and tourism to a new generation of predators, made possible by the Internet and availability of cheap flights to Southeast Asia (ECPAT International, 2016).

In the wake of the Atlanta shootings, debates over whether or not the women victims were sex workers served as a punctual reminder of the routine intersection of race, gender and sex work for ESEA women. There were those who assumed the women must be sex workers simply because they were Asian and perceived to be working class, those who argued that (sex worker or not) speculation was culturally shameful and therefore disrespectful to the memories of the women, and finally, those who hoped to create a dialogue on the expansive harm of stigmatising sex work among migrant women workers.

Yves Song, a member of the abolitionist grassroots group Red Canary Song, which operates in the tradition of migrant sex worker mutual aid, explained that, though it was important not to make assumptions, whether or not these women were sex workers did not escape the reality that, firstly, many migrant women *do* rely on sex work for survival, and secondly, the shooter believed that they *were* sex workers, which identified them as targets (Nguyen, 2021).

VILLAIN OR VICTIM? WHEN BETWEEN BORDERS

Bordering practices and the policing of migration have long been used to ensure the experiences of 'belonging' remain liminal for racialised people in the UK (see Chapter 3: Migration patterns). For ESEA men, this has often been reinforced

through racialised and sexualised perceptions of desirability –
both hypersexualisation and desexualisation have served as
tools of exclusion.

Racialised fears surrounding Chinese male labourers were
fueled by political exploitation and popular fiction (see Chapter 3:
Places: Where we gather). Through a gendered lens, Chinese
men were cast as both economic and sexual threats, with
hypersexualised narratives – such as Sax Rohmer's Fu Manchu
thrillers – depicting sinister 'Chinamen' luring white women
into opium dens. Furthermore, the 'Compulsory repatriation
of undesirable Chinese seamen' issued by the Home Office in
1945, a state-enforced removal, had lasting, gendered repercussions (LJMU, 2023). The secret deportations fractured
families, leaving wives in social and financial precarity and
children with the absence of paternal figures (see Chapter 3:
History books, but for whose history?).

For the men who remained in the UK, racial anxieties
around ESEA masculinity did not disappear; they evolved.
The heartbreak of involuntary separation made way for new
forms of gendered exclusion. Stripped of traditional family
structures, they were framed as men without roles – neither
fully accepted into British masculinity nor granted the space
to form their own. Accusations of asexuality or homosexuality
further alienated them. At the same time, they endured ongoing hostility and were often pushed into feminised labour such
as cooking and laundering, jobs dismissed as 'women's work'
that deepened associations between ESEA men and sexual and
gender deviance.

Excluded from the nuclear white family and erased from
dominant images of British identity, ESEA men were posi-
tioned as outsiders in both public and private life. Even today,

racialised exclusions persist in more insidious ways – it is telling that contemporary AI image generators struggle to depict an Asian man with a white woman (Sato, 2024). If the hyper-sexualisation of ESEA men was once an expression of white male insecurity, their later desexualisation was an attempt to soothe it.

Just as ESEA men have been historically excluded from dominant family structures, migrant ESEA women, particularly those with lower economic status, are often relegated to pre-carious roles that reinforce colonial-era gender dynamics such as through marriage, labour, or the criminalisation of their work, perpetuating global inequalities and exploitation. Some women enter relationships with men in the Global North through mail-order bride services, creating vulnerabilities to abuse, since their visas and recourse to public services are dependent on their marital status. Meanwhile, other migrant women are tied to domestic labour, which has been made more precarious under the Hostile Environment, and sex work, the criminalisation of which creates easy grounds for exploita-tion (see Chapter 6: Workers' rights in the age of the Hostile Environment).

While anti-trafficking policies claim to protect women, they often conflate migration with trafficking, leading to border crackdowns that harm sex workers and criminalise voluntary migration, as well as increasing racial profiling among Asian, Black and Latina women sex workers. Between 2012 and 2016, due to a crackdown on massage parlours, the number of Asian people charged by the NYPD with 'unlicensed mas-sage' skyrocketed by 2,700% (Mac & Smith, 2018). The way that legitimate concerns over trafficking are lumped in with the broad criminalisation of sex work ignores the larger issues of undocumented migration and exploitation, which almost always stem from systemic issues rather than individual ones.

Mac and Smith also argue that many of the western NGOs who focus on trafficking tend to be religious, or led by carceral feminists (or both), and that their focus is almost always on trafficking into prostitution and the subsequent need to criminalise and abolish it, rather than on improving material conditions and granting agency to the migrant sex workers themselves.

BELONGING, PARTICIPATION AND DISSOCIATION

If I walk down the intersection between Soho and Chinatown,

Where do I belong?

For if I should belong in both,

Do I belong in neither...

(Zhang, 2021)

When exploring the relationship between Soho and Chinatown, poet SJ Zhang highlights the dilemma faced by individuals who are both ESEA and queer, while challenging gay, white poetry that idealises Soho as the final destination for queer liberation.

The existence of the phrase 'no fats, no femmes, no Asians' is telling enough about the blatant racism and discrimination that ESEA individuals can experience in queer spaces, especially online, where anonymity is afforded, whereas homosexuality is generally still an extremely taboo topic across ESEA cultures despite a (hidden) history of LGBTQ+ art, literature and practices

(Nugroho, 2016; Samshasha, 2015). After all, queerness is inherently associated with the Western concept of individuality, in contrast to collectivist and family-oriented ESEA cultures.

This conflict unfortunately means LGBTQ+ ESEA folks feel they have to choose between the queer identity they have come to embrace and the family values instilled in them. 'How can you be proud to be East Asian if you've turned your back on your family which goes against tradition?', laments trans activist Eva Echo (2023) when navigating feelings of pride on a political level but guilt and shame on a personal one. Elsewhere, we witness in China a growing 'yes-but-not-yet' culture (Luo et al., 2023) amongst queer youths, to afford themselves a queer life with an expiry date, recognising the inevitability of facing the pressures to conform and start a traditional family. While concrete figures don't currently exist to indicate that LGBTQ+ individuals from racialised backgrounds are disproportionately affected by some degree of familial rejection in relation to their sexuality or gender identity, we can look to a study by the HRC Foundation revealing only 17% of AAPI LGBTQ+ youth say they can definitely be themselves at home (Kozuch, 2019).

Family, ancestry, lineage, heritage, legacy, inheritance… These are terms traditionally associated with the patrilineal system and heterosexual reproduction. Where do queer people, especially queer ESEA people, fit in? Favouritism towards the male child, the firstborn in particular, is the default power dynamic in most traditional ESEA households, following three Confucian codes, which 'regulate male behaviour and… privilege men in governing the family as fathers and ruling the nation as emperor and bureaucrats' (Cheng, 2016); filial piety toward parents and ancestors (xiào 孝), brotherhood (tì 悌),

and loyalty (zhōng 忠). This isn't limited to families that subscribe to Confucian values, as Muslim, Christian and Hindu practices also echo the same social conservatism with varying degrees of slating homosexuality as immoral. In this context, being queer, in particular a gay man or a trans woman, is a failure, a refusal to participate in society.

Where does one go when even home may not feel like an option?

The dichotomy between life at home and in society is further highlighted when observing the pressures of patrilineality from the lens of ESEA masculinity. In an interview featured in Giboom Park's *Not Your Yellow Fantasy* (2020), Zhang, a Chinese American, shared:

> It was just weird to constantly be spoon-fed this toxic masculinity all my life by my family where I had to 'be a man,' but the minute I stepped foot outside my house, it seemed like no one was going to 'see me like a man.' If you wanted to be treated like a man, whatever that means, you had to be in an Asian circle.

It can be particularly destabilising to bounce between ridicule and reverence, a double bind that can pressure ESEA men to either reassert power by further oppressing others within the household, or more commonly, to save face and resort to dissociation under the guise of maintaining stability and resilience as an exemplary ESEA trait. TV presenter Frankie Vu (2023) reflects upon chịu khó (忍𠚮), a Vietnamese term his father instilled in him to endure hardships by 'getting on with it and not talking about it'.

By this logic, some might choose to even distance themselves from other members of ESEA communities as a way of assimilation, or as a means of self-protection against racial discrimination. How many meaningful potential friendships have been overlooked because of this? Alongside the social and cultural stigmas around expressing vulnerability and risking stability, this self-protection logic may also deter ESEA men from engaging in activism. Second generation Indian Malaysian, gender non-conforming writer and performance artist, Alok V. Menon (2024) has described a vicious cycle of invisibility and dissociation, in which increased visibility is linked to fears of increased discrimination:

> We perfected dissociation at a community level. Our trauma coping strategies become our cultural identity. When you come from a heritage of intergenerational shame, you grow up not being able to tell the difference between a body and an apology.

AGAINST THE WHITE MEDIAN

When challenging stereotypes of ESEA men and women, we must carefully consider whether the alternative representations we propose remain constrained by standards imposed by white supremacy. For example, when we argue that ESEA women should not be eroticised, are we inadvertently reinforcing white feminist ideals of female purity? Similarly, when ESEA men are celebrated for their muscularity or marketability as seen in the rise of the 'Asian Himbo'[1], do we risk perpetuating a limited and stereotyped notion of masculinity, shaped by white supremacist standards of desirability?

---

[1] A hunky, Asian 'bimbo'.

Historically, exclusion has consistently resulted from these gendered perceptions, regardless of which direction they lean. While Black and Brown men have been demonised as sex-crazed animals, posing a danger to (white) women, the Asian Himbo trope often portrays a lack of intellect, suggesting they can only have brains or brawn. Meanwhile, white European men occupy the perfect middle ground of masculinity and sexual appeal in an attempt of race realism to uphold traditional constructions of hegemonic masculinity (Park, 2020).

The implicit coupling of men with masculinities that are present in many critical men's studies ignores the complex intersections between sex and gender. Take the group BTS as an example. They've introduced a new form of versatile masculinity, positioning themselves as a multinational, multicultural brand whilst retaining national specificity by embracing Pan East Asian concepts of 'soft masculinity' (such as *bishōnen* in Japan and *hwarang* in Korea) alongside Western ideas like metrosexuality. It is a cultural priority to present beautifully in Korea regardless of gender (Morin, 2020), proving that gender presentation does not map onto sexuality in any way. Despite homophobia still being very common, Korea is a homosocial society where same-sex closeness is common among men without implying sexual relations.

Female masculinity is a particularly fruitful site of investigation because it has been vilified by heterosexist and feminist/womanist programmes alike; unlike male femininity, which fulfils a kind of ritual function in male homosocial cultures, female masculinity is generally received by hetero- and homonormative cultures as a pathological sign of misidentification and maladjustment, as a longing to be and to have a power that is always just out of reach. Within a lesbian context, female masculinity has been situated as the place where patriarchy goes to work on the female psyche and reproduces misogyny within femaleness. By exploring female masculinities, we can understand how masculinity itself is constructed and questioned, shedding light on its complexities beyond traditional norms. After all, masculinity becomes legible as masculinity where and when it leaves the white male middle-class body (Halberstam, 1998).

This is why we need to encourage a total upheaval of our cultural viewpoint away from whiteness as rightness, exploring gender beyond traditional western frameworks of masculinity and femininity. We must break free from the oppressive gender binary which intersects with other power structures to perpetuate material oppression worldwide.

QUEERING PERSPECTIVES

Queer perspectives are at the vanguard of efforts to challenge traditional representations of every sexist society, highlighting the diverse experiences and identities within LGBTQ+ communities and beyond. They confront ingrained stereotypes and advocate for greater inclusivity, reshaping narratives to reflect the richness and complexity of queer lives. These redefined narratives not only validate queer experiences but also offer alternative frameworks for understanding gender, power and resistance – challenging the structures that confine all those living under patriarchy. As we recognise gender as diverse and fluid, we choose to destabilise cis-heteronormativity and dismantle compulsory heterosexuality. Responses to a survey on queerness by online platform daikon* zine (2018) echo this sentiment about destabilisation:

Queerness for me is the excitement and fear that my body doesn't do what it 'should' do. It is the joy and confusion of often wanting my body to be other than it is. It is delight and shame at upsetting binary heteronormativity.

This destabilisation extends towards how we frame time. Jack Halberstam's concept of Queer Temporality (2005) challenges the idea that queer lives follow a linear, heteronormative time-line. Milestones like marriage and children, central to hetero-normative success, may not hold the same meaning for queer people. Queer Temporalities are especially true for transgender individuals undergoing medical transition, such as Hormone Replacement Therapy, which can trigger a second puberty at any age. Success and fulfillment for LGBTQIA+ individuals, therefore, may not align with mainstream cultural expecta-tions. Queer Temporalities question how marriage, genera-tivity and inheritance shape ideas of maturity and happiness (Goltz, 2009).

Moving away from reinforcing homonormativity – the replication of heterosexual norms within queer spaces – is essential. This includes rejecting exclusionary attitudes from SWERFs (Sex Worker Exclusionary Radical Feminists) and TERFs (Trans-Exclusionary Radical Feminists). Queerness resists rigid structures, allowing space for growth and change. The idea that queerness might be 'just a phase' is often dismissive, yet it reflects the truth that identities – gender, sexuality and rela-tionship structures – naturally shift over time. 'Phases' are not trivial; rather, they are stages of clarity and self-acceptance, punctuated by necessary periods of confusion.

Nuallak (2022) encourages us to embrace multiplicity beyond the gender ternary of man, woman and an intermediary that negotiates between the two, and to 'recognise the incoherence of non-binary so we can agitate for our diverse and varying needs and desires'. By allowing ourselves the permission for fluidity and incoherence, we create space for imagination and disorientation, freed from the urgency to label. Musician Jason Kwan (J. Kwan, personal communication, June 2023) echoes this sentiment, sharing a reminder not to 'pressure yourself to figure out exactly who you are, because by tomorrow, you could be someone completely different.'

Feminist and queer theorists argue that sexual and gender identities are not merely innate but shaped by agency and self-discovery, in particular the concept of performativity and gender as a series of repeated performances (Butler, 1990). Moving beyond the belief that preferences are fixed or predetermined allows for greater ownership and accountability over one's evolving identity. For ESEA individuals, the ambiguous nature of sexuality can be especially empowering. Take this response from daikon* zine's survey on queerness (2018):

> Queerness to me legitimises how I experience attraction to people in ways that have changed over time and are influenced by my experiences and understandings of myself. It allows me to be uncategorisable which gives me a feeling that I have more agency in my sexuality, than more societally visible aspects of my identity like the ways I'm racialised and gendered.

To actively resist categorisation and engage in persistent desire is to recognise that we have within us power and possibility. And this feels more necessary than ever when one is marginalised.

Khai Ramli, creative practitioner and community organiser, shared her empowering experience of exploring the intersection of gender presentation and religion through wearing the hijab:

[It] is a new phase of my life... To me, it's spiritually to better myself... During summer, [I can sense that] a lot of people look at me like I'm oppressed... This is my choice. Wearing a hijab is just part of my growth. (N. K. Ramli & F. Abdullah, personal communication, 15 June 2023)

Queerness as a framework encourages us all to cultivate a caring and inquisitive relationship not only with our body-mind, but the connections it fosters to others. It instils in us a sense of openness by default in recognition that we all have the capacity for change and fluidity. By embracing this, we can engage with others from a deeper place of empathy and unconditional love, releasing the expectations and pressures that often weigh us down. It teaches us to let go of pre-conceived ideas about who we 'should' be, and instead be present and attuned to who we are becoming.

Queerness allows us to move away from the fragmenta-tion of identity that arises from the tension between ESEA cultures versus gender expression and sexuality in the guise of Western cultures. By embracing queerness and rejecting white dominance, we also challenge the notion that there is a 'right' way to exist, love, or be. It teaches us that we are not confined to predetermined paths of desire, identity or expres-sion. Ultimately, queerness as a framework invites us to honor the complexity of all identities, allowing for the possibility of change, transformation and growth. As we resist categorisa-tion, embrace fluidity and prioritise empathy, we create space for a more inclusive, dynamic understanding of ESEA gender and sexuality that celebrates the richness of our experiences and the complexity of our identities.

TOWARDS AUTONOMY AND SOVEREIGNTY

This chapter has raised questions over many unexplored territories. These include deeper explorations of how religious ESEA experiences are shaped by their gender; wider considerations of queer temporality and intergenerational communication on sexuality; receptivity of feminist movements in the UK in comparison to Asia; newer feminist interpretations of prevalent ESEA philosophies like Confucianism; stereotypes of Asian mothers and aunties, and when gender intersects with parenting through the ESEA lens.

Embracing queer perspectives could be the answer in cultivating humanity and meaningful connections from a place of openness, where we can honour the sovereignty of the body with respect. A lens that could also embolden ESEA individuals towards unapologetic new expressions of self.

At the same time, we embrace and encourage attempts to disrupt the heteronormative, eurocentric status quo that places expectations on ESEA people's bodies, sexualities, and gender roles. Authors like Nguyễn Phan Quế Mai are challenging the dominance of American narratives in English language books about Vietnam, giving voice in her fiction writing to the lives of Vietnamese women and, in the case of *Dust Child* (Nguyễn Phan, 2023), particularly sex workers, whose experiences have rarely been considered. In recent years, groups like the Sex Workers Advocacy and Resistance Movement (SWARM) in the UK have made attempts to recentre and reclaim discourses on sex work and migrant bodies through awareness raising and mutual aid building.

In creative spaces, groups like ESEA Life Drawing aim to challenge taboos in many ESEA cultures around self-expression, nudity, sexuality and bodily autonomy, as well as the social narrative that assumes ESEA women to be demure, meek and submissive. Meanwhile, on stage, Kimber Lee's ferocious play, *untitled f*ck miss saigon*, premiered at the Royal Exchange Theatre, Manchester, in 2023, which threw down the gauntlet to white audiences over their expectations of Asian women in theatre narratives.

These efforts demonstrate that change is happening, albeit slowly, and that the work of deconstructing harmful stereotypes and reclaiming agency is an ongoing process. As we continue to explore these uncharted territories, it is crucial to engage in dialogue that not only challenges normative gender expressions and sexualities but also empowers ESEA individuals to own their experiences. This broadens our thresholds of desirability and self-acceptance to move beyond the restrictive confines of a white supremacy framework. Through collaboration, openness, and persistence, we can dismantle the rigid frameworks that limit ESEA people's gender and sexual autonomy, creating a new blueprint for liberation.

8

ORGANISING THROUGH CRISIS

THE SYSTEMS AT PLAY

Activism and social organising are nothing new, but the way we organise has changed radically since the early 2000s with the proliferation of social media. Even as a group, our ways of thinking, being and organising in the world as besea.n have transformed in the space of just five years since our founding. Much of this chapter is shaped by the co-authors' own experiences, and the challenges detailed within speak to an important part of our growth as a group, though the views expressed are our own. This chapter is an effort to embrace the struggles that come with the work, in the hope that it gives us clarity and hope for the future.

There is no denying that mass organising in the past fifteen years has been vastly shaped and aided by the rise of social media and decentralised or 'horizontal'[1] types of organising.

---

[1]An ideology that privileges lack of hierarchy or leadership, where decisions cannot be made by anyone for anyone else.

This is often in response to a catalysing, viral image or video that indicates abuses on the part of a state or institution, which sparks outrage and mass mobilisation across various means and platforms. An obvious example is Black Lives Matter, but we have seen similar mobilisations in what Western media called the 'Arab Spring' in the early 2010s. While we do not have the scope to compare and contrast various global struggles and political mobilisations in detail, and make no attempt to do so, we can draw from the work of others and recognise the dangers of virality in an era of rapid, digital information sharing, where inaccuracies and premature conclusions are more prolific than ever. Although our own group was founded and grown through Instagram, we know all too well the risks associated with widespread digital responses to offline problems.

Of course, not all activism has to be part of a mass movement, and not all community builders will view their work as activism. However, we believe in the critique and dismantling of systems rather than holding individual people or small groups of people solely responsible for – in some cases – centuries of discriminatory beliefs and practices. That's not to absolve individuals of their moral responsibilities, of course, but when dealing with the tricky subjects of race, class, and environmental, queer and disability justice, emotions tend to run high. It's easy to get caught up in interpersonal struggle without looking at the bigger picture. One person is not representative of an entire system, and the most effective strategy for change should not and cannot focus on that one person alone.

This may be an easier concept to grasp when we think of individuals we may know personally, but it's more difficult to see when it comes to our political leaders. To be clear, a country's leader may be associated with various policy changes or party behaviours that impact the lives of millions, and they should absolutely be held accountable for their actions. But representatives come and go. Establishments and the ideologies deeply embedded within them last a lot longer. Government terms are short and limited in their scope for real change, so groups implicated in action for policy change must be prepared to weather the seasons.

Tackling the interconnected issues of racial capitalism, environmental justice, poverty, ableism and many other struggles will require a cross-community, multi-pronged approach that puts service to communities, rather than profit, at its core. It is a daunting, long-term work that requires action within, beyond and against systems – and one that may require us to look beyond the scope of our own lifetimes. The framing of these different, interconnected struggles against systemic oppression as a grounds for overthrowing those systems, is often referred to as a 'liberation movement', or simply 'liberation'. Quite simply, it imagines a transformative global and social change, and freedom from oppression and marginalisation for all people. We will use the term in this context throughout this chapter.

The concept of decolonisation – particularly the decolonisation of knowledge systems – is something that we hear a lot these days. There are ongoing movements and debates around decolonising education, museums, the outdoors, universities and global health, all of which root traditional ('Western') systems of knowledge in colonialism. Colonial ideas are an essential tool for the everyday functioning and survival of growth-focused capitalist systems, which contribute to widening inequality between the richest 1% and the rest of the world. Decolonial approaches question and challenge eurocentric histories of power (through thought and practice), as well as acknowledging and confronting the wide-ranging impacts of empire and colonialism – including their influence on global economies and labour markets.

Colonialism may be perceived as something confined to the past, but its legacies remain embedded in the ways Global North countries interact with – and often exert influence over – the Global South. This is evident in government and institutional intervention (such as military involvement or projects through the World Bank or the UN) as well as in individual actions (aid workers, journalists, consultants and other individuals with the relative freedom to navigate Global South spaces on their own terms). Furthermore, particularly in the case of the UK, colonialism played a central role in shaping the country's present day wealth and economic status.

It is easy to overlook the connection between colonialism and the social issues that persist through the contemporary 'culture wars' that dominate our headlines. In 2024, after the violent mass stabbings in Southport that resulted in the murders of three children and left ten injured, a media scapegoating campaign led riots to erupt across the country. While the racist and violent actions of many individuals involved in these riots – targeting racialised people of all ethnicities – were widely condemned, social and political commentary has pointed to the general collapse of public services under previous governments, along with an over-focus on culture wars. These factors have contributed to widespread inequality and disillusionment, which serve as key undercurrents of nationalist extremist acts among overwhelmingly working class people. The wedge driven between white British working class populations and all other working class populations is strongly linked to colonialism.

Part of the negative myth-making discourse around migrants is that they are afforded privileges, including state welfare, not afforded to citizens. While this is far from the case, academics like Dr. Nadine El Enany also argue that the country has not yet faced up to the relationship between colonialism and *all* white populations, including the working class.

According to El-Enany, it is Britain's collective 'profound colonial amnesia and imperial ambition' (El-Enany, 2020) that has resulted in the damaging and dangerous construction of a post-Brexit society, with an ever widening gap between migrants and the local population. Furthermore, as immigrant communities in the UK have 'integrated' into society over generations, it is easy to forget that the issues facing working class people, who pay the highest price under extractive and exploitative systems, are often shared across racial and ethnic lines; especially as exploitation has taken new forms in our modern capitalist society.

Fundamentally, while money from colonialism *was* funnelled into British society and infrastructure, the fruits of that project – both at the time, and in the present day – were designed to benefit the elite, and the repercussions target both white and racialised working class people alike. The problem is not that white British working classes yield colonial privilege over immigrants, nor that immigrants are taking over the country and all its resources: the problem is that we have forgotten who the real enemy is. Thinkers like El-Enany have posited that the scapegoating of migrants is a strategic move on the part of the elites who make up the UK establishment, which is unwilling to face accountability for its colonial past.

Under capitalism, society functions around commodity production, wherein value is created by the worker and extracted by the owners of the means of production, or of the capital it generates. Under racial capitalism, the system has developed and continues to function through the disproportionate exploitation of racialised and marginalised people. These days, particularly in the UK, daily life has become increasingly commodified, and immense power is concentrated in the hands of a small percentage of very rich individuals. Consumerism has found its way into parts of our lives we may not even register: practices for mental and physical health have become commodified through the so-called 'self care' economy, the fitness world relies on a neverending consumption of goods and services, as do privatised healthcare and pharmaceutical industries.

Economist Christine Berry has claimed that overwhelmingly, it is wealthy white people in the UK who have benefited from the colonial practices that turned assets, land and people in former colonies into financial assets in the City of London. This has led to the exploitation of the working class not just as workers, but as tenants, debtors and even healthcare patients (Berry, 2021). Any attempts to truly decolonise thought and action must start, therefore, with recognising and understanding the workings of class inequality, colonialism and how both – alongside other intersecting systems of oppression – have structured society. This includes the participation of migrant communities within that system, as well as the ongoing impacts of imperialism and capitalism on marginalised groups worldwide. We have learned to separate different types of racism, condemning acts of extremist, nationalist violence as fringe behaviour, along with a healthy dose of classism and ableism, with many social media commentators criticising rioters' presumed lack of education or access to public health services. Yet we fail to interrogate their connection to colonialism and capitalism.

Approaching decolonial ways of thinking and doing is an immense task, and one which can be broken down into multitudes of areas and disciplines. We recall here Dr. Ruha Benjamin's encouragement for every person to find and tend their own 'plot', or quite simply, everyone has a role to play in movements for change (see Chapter 2: A purpose beyond a label). What we hope to do in this chapter is to understand, critique and propose new avenues for current advocacy, activism and organising efforts, with a particular focus on digital organising, as well as to shine a light on ESEA activism and resistance movements, past and present.

The ideas we lay out below are the result of years of collective organising among ourselves, our peers and other communities we respect and admire, as well as invaluable ideology and social commentary from various writers and thinkers. They are strongly informed by an anti-capitalist, Black feminist lens, drawing on ideas from abolitionist[2] practitioners and solidarity economics[3]. They are not prescriptive doctrines; rather, we hope to encourage alternative ways of working and moving together through crisis. We may not always agree with each other, but through

---

[2]The political movement to abolish carceral systems such as the prison industrial complex, policing and detention, while actively building new structures that promote safety, care and justice.

[3]Solidarity economics is an approach to organising economic activity based on cooperation, mutual aid and collective wellbeing, rather than competition and profit maximisation. It prioritises people over profit, with an emphasis on democratic decision-making, shared resources, and social and environmental justice.

mutual respect and constructive debate, we hope to play a meaningful part in the fashioning of a fairer society.

ADVOCACY AND ACTIVISM

Advocacy and activism are two words we often hear used interchangeably, but it's important to know the difference between the two. Advocacy is a form of support for particular causes, which often involves amplification of and education on those issues, while activism tends to involve direct action to bring about social, political, economic or environmental change. Advocacy becomes activism when it takes a particular form in the public domain, and it is possible to be both an advocate and an activist at the same time. Advocacy and activism can take place at an introspective level (challenging one's own biases through unlearning and relearning), interpersonal level (through social interactions that encourage people to engage), institutional (working to effect change within a school, workplace, union or other institution) and structural or systemic (working to effect change across multiple axes in order to precipitate systems change – anti-capitalist organising is a good example of this). However, shifts away from activism towards advocacy at an institutional and NGO level have hampered grassroots activism in recent decades, as we will explore later in this chapter.

There is a common perception that activism is the remit of young people. Indeed, movements such as Fridays for Future, Palestinian Youth Movement and Dear Asian Youth have been important movers and actors in environmental and liberation organising in recent years. And yet some of the most committed activists we have seen are from older generations: these people have often been steadfast and consistent in their commitment to resistance against oppressive systems, and they have a wealth of experience to share. While youth activism is a critical lever in the mechanism for public education, particularly in the digital age, these movements must rely on tangible, generative exchanges with older generations.

We believe there are also valuable insights on offer by applying the principles of practices common in many ESEA cultures (and other cultures worldwide) of venerating ancestors, honouring elders and passing on collective community wisdom through traditional practice and storytelling. In this way, solidarity practices become a commitment that endures beyond an individual life and well into the future. Such was the approach discussed at besea.n's 2023 ESEA Heritage Month event at the British Library, entitled 'Roots of Change', which saw the host, panellist and audience members explore ways in which we can take the ideas and principles of ancient practices and ground them in intergenerational knowledge exchange, in order to advocate for environmental liberation.

FROM THE STREETS TO THE SCREEN

There is no doubt that social media has radically changed the way much grassroots advocacy and activism are carried out. The Black Lives Matter movement, which has been active since the early 2010s, gained explosive, global recognition after the killing of George Floyd in 2020. Activists took to social media using the hashtag #BlackLivesMatter. Within hours, a combination of global outrage and social media algorithms propelled the subsequently organised protests, marches, panels, healing circles and fundraising efforts into the world spotlight.

Further still, the movement sparked interrogation and investigation into practices of systemic discrimination across public institutions, companies and personal interactions worldwide. Though, at the time of writing, political developments in the USA regarding affirmative action and other inclusive policies – including the threat of further backsliding under a second Trump administration (for example, the Executive Order eliminating affirmative action for federal contractors) – make it hard to see how much meaningful change has really occurred at an institutional level.

Other movements that have gained global traction through use of hashtags include #MeToo, #WeAreNotAVirus and #StopAsianHate. In the UK, against the backdrop of rising pandemic-related racism, a Masterchef contestant caused widespread anger when she indicated that ESEA cuisines are 'dirty'; a common trope that suggests the unhealthiness of ESEA foods, as well as the lack of value in both the ingredients and the labour behind the dishes.

Community organiser Anna Chan and podcaster Georgie Ma, both of whom grew up living and working in their parents' Chinese takeaways, came up with the hashtag #ESEAEats: a counter-narrative to the 'dirty' trope, encouraging people to share their experiences and stories about the food of their heritage (besea.n, 2020). Hundreds of people took to Instagram – and what risked fizzling out as a collective social media rant actually transformed into an opportunity for connection between different actors in the food industry.

Identity was a focal point of these connections, with many ESEA chefs, 'takeaway kids' and food writers drawing on their heritage to spark important interrogations on authenticity, disposability, homogeneity and the ways in which cuisines are hierarchical and Othered. Many of these conversations made efforts to recognise the connections to colonialism and racism, asking why it is that foie gras and blood sausage are considered culinary delights, while it is often immediately assumed that people from ESEA cultures eat dog meat – the barbarism being heavily implied. #ESEAEats teaches us how much healing and intercultural connection is possible when we break bread together.

There are drawbacks associated with hashtag activism, of course. In the case of the Black Lives Matter protests of 2020, the accompanying 'black square posts', initially in solidarity with Black musicians, involved Instagram users posting a black square on their grid and using the #BlackLivesMatter hashtag. More and more people began sharing the hashtag simply in support of Black lives, with little context or understanding of the initial movement. As the algorithm did its work, the posts of local organisers – attempting to share important information about protests and action items – were drowned out. Activists called for people to use the alternative hashtag #BlackoutTuesday instead, but the damage was done.

Furthermore, as Mariah L. Wellman (2002) has described,

> [The] posting of black squares was performative allyship utilised [sic] strategically to build and maintain credibility with followers. Influencers were unable to genuinely merge their existing brand image with the Black Lives Matter movement long-term, resulting in the meme-ification of social justice activism and no substantial progress toward diversity, equity, and inclusion within the wellness creator industry on Instagram.

While it's important to recognise the role of social media in spreading awareness and maximising engagement, especially in a society where we increasingly understand popular culture and identity through short form content and memes, it's critical to translate online clicks to offline action. What good is sharing a post if no action is taken to change behaviours and policies? How can we protect vulnerable people who may get caught in the backlash or crossfire of posts shared virally with little context or nuance? How can we ever condense real organising principles, those needed to grow and build a movement, into 280 characters?

In addition to the concept of performativity, in what is increasingly known as the 'attention economy' (the monetisation of human attention online), engagement with social media activism is vulnerable to narcissism and self-promotion, even at unconscious levels. When social media messaging is structured and promoted around the self – and often, as part of, or a complement to, one's 'personal brand' (since people themselves have now become a commodity online), it is easy to fall into the trap of sharing – or, indeed, oversharing – as a response to guilt, trauma or burnout. Widespread, viral socio-political movements often have the effect of making us question our own beliefs and behaviours, particularly those focused on systems of oppression such as racial capitalism or environmental colonialism.

In a world where many people tend to document every thought process online, admissions of guilt – about one's actions, or, indeed, inaction – are often couched in social justice language, disguised as advocacy. Excessive guilt-posting, when done at scale, often means that the affected communities are sidelined or overlooked, and the initial message becomes secondary; the particular digital space is essentially saturated. This is especially the case if the person is a celebrity or influencer.

Worse still, because of the way the attention economy functions, if a post is well-received and garners a lot of likes, shares and saves, such activity presents a favourable business opportunity to both individuals and corporations.

For proof of this, we need only to look at the ways in which movements like Pride and Black History Month have been co-opted and absorbed into corporate PR exercises, much to the dismay of many individuals and community organisers. As DEI practices increasingly become more about branding and corporate box-checking, the appropriation further serves to obscure genuine worker organising to secure better rights, fairer treatment and better working environments for racialised people. In short, 'diversification' within the system blocks out actual attempts to either neutralise the harms of the system, or do away with it all together.

For community members themselves, the online world is just as fraught and fractured as physical diasporic communities. When we say 'community', we are referring to an imagined ESEA online community – there is no one single ESEA 'we'. As we've seen, the short form, rapid sharing nature of most social media content means there is little space for nuance. The culture of 'calling out' online, though often well-intentioned, especially when it comes to organisations or institutions, risks being less about actually addressing change with the targeted individual or group of individuals, and more about staging a conflict for an audience.

While condensing complex social and political issues into short, bulleted infographics or captions can make the overwhelming world of politics and current affairs more accessible, there is a risk that individuals will use social media, limited as it is, as their only source of information and/or connection-building, rather than as a complement to more nuanced, complex inter-rogations, reporting and organising. There is a strong risk for people to reserve their battles for online spaces where they are afforded the protection of a screen, without taking any actions with them into their offline world.

Sometimes, though, this environment – the online arena – can lead to a polarisation of opinions that bleed into every part of our lives. In such cases, our online 'personas' take over and erase nuance, explained writer and researcher Vy-Liam Ng in an interview (V.-L. Ng, personal communication, 7 August 2023). 'Our online personalities are kind of our avatars for this public, political battleground,' Ng explained, referring to offline con-versations with loved ones, in which these political 'avatars' make our conversations rigid, unempathetic and unlikely to have any real impact. This can lead to a failure to convince the receiving 'opponent', who is likely to reject being moralised, and frustration and disillusionment on the part of activists or advocates, as they do not see their work online connecting to any real-life practical application.

Online spaces can often lead to an intense focus on uni-form thinking, exclusion and hyper-reactivity (staying on activist 'trends'), which can lead to burnout or, ultimately, in extreme cases, push people into polar opposite political views.

When we are consumed by online action, it is easy to for-get about those who do not live predominantly online or stay active in digital organising circuits. In the case of ESEA com-munities, this might be older, first-generation people, those without digital access or literacy, or vulnerable migrants work-ing in labour conditions that don't afford them the time or energy for such online engagement. Or, simply, there are also many who have chosen to forgo the social media life, for the betterment of their mental wellbeing, or to reduce participation in systems that are increasingly dictated by tech oligarchs. The imagined audience, therefore, often embodies the same persona as the person posting content; and thus the echo chamber is created and perpetuated.

ACTIVISM AND ESEA COMMUNITIES, HISTORICALLY

It is an easy assumption to make that ESEA communities in the UK are not overly political or engaged in activism of any kind. These assumptions tend to stem from a generic understanding of cultural and generational practices, which risk leaning into the familiar yet consistently perturbing model minority myth stereotypes. Or perhaps this perceived non-participation is the result of social exclusion. There are, however, long-flowing currents of advocacy and activism among many ESEA communities in our society, even if they did not or do not refer to themselves as such. To our knowledge, few attempts have been made to join the dots. This book is not an ethnography, nor an in-depth study of the history of community organising, but it does make some attempt to introduce a subject that is mainly understood in activism and/or academic spaces to a wider audience.

While there have been public awareness raising and anti-racism campaigns led by ESEA individuals and groups impacted by racism during the COVID-19 pandemic, these efforts are part of a long history of activism and community organising, reflecting the enduring spirit of resistance among ESEA communities. The Chinese seamen in Liverpool, who were secretly deported by the UK government (see Chapter 3:

History books, but for whose history?), went on strike in 1942 over pay discrepancies with British workers, represented by the Liverpool Chinese Seamen's Union. They won a pay increase of £2 a month, and the right to the standard £10 a month 'war risk' bonus (Oyen, 2013).

In the early 2000s, Chinese restaurant owners and workers pushed back against the scapegoating of the Chinese restaurant trade by the British government and media for the foot and mouth disease outbreak, with protests taking place in London and Manchester Chinatowns. Jabez Lam, a community leader whose work spans decades in the organising space, was supporting Chinese victims of hate incidents at the time. He reported an increase in reports of vandalism, violent attacks and restaurant boycotts. Overall, Chinese businesses across the country reported a 40% downturn in trade. In response, Lam, member of the popular website Dimsum and various Chinese community groups, such as the London Chinatown Association, formed the Chinese Civil Rights Action Group (CCRAG) and set out to resist the unfounded allegations and protest the racist treatment of the workers. A media and communications campaign was launched, with a call to action gathering protesters in London, which saw a turnout of over 1,000 people (Soh, 2020).

Resistance against yellowface practices and the historical oversight of ESEA actors in the arts, which has significant impact on the livelihoods and wellbeing of community members, has been a regular target of activist groups and politically engaged production companies such as New Earth (formerly Yellow Earth), BEATS and Moongate Productions. Some examples include the protest movement against the Royal Shakespeare Company for its failure to include sufficient Chinese or East

Asian actors in its 2012 production of *The Orphan of Zhao*, and a 2019 letter, organised by BEATS, signed by over 100 ESEA people in the film and television industry to call for greater diversity and fairer inclusion for screenwriters in the CBBC sit-com *Living With the Lams*.

Organising movements have long been engaged in the fight for racial justice, particularly with a family-centred approach that prioritises support for those impacted and their family members. Lam, presently an organiser with ESEA Community Centre (formerly Hackney Chinese Community Centre), has been an active member of The Monitoring Group (TMG), one of the oldest grassroots anti-racism organisations in the UK. Born from the experiences of African Caribbean and Asian people who grew up in Britain during the 1970s, TMG has since played a significant role in the shaping of legislation and social practices in the UK, and been at the forefront of family-centred campaigning for racial justice.

Some of the causes they have spearheaded over the last few decades include the New Diamond 5 campaign to support Chinese victims of an attack in London's Chinatown and support for the families of the Dover 58, a group of Chinese nationals who were tragically killed in a smuggling incident. The group also led a shutdown of London's Chinatown in protest against a series of 2013 immigration raids found to be motivated by racial profiling rather than actual intelligence (The Monitoring Group, n.d.).

In addition to organising in the streets, online engagement spaces have also sought to tackle issues of misrepresentation and erasure in cultural institutions, such as the website British Chinese Online's campaign to address the British Library's refusal to engage with the topic of the opium trade in a 2002 exhibition on the British East India Company. The mobilisations organised and supported through websites like British Chinese Online and Dimsum, and through affiliated groups like CCRAG or Min Quan, may be the first time that many users have engaged in any political action. In this way, the digital spaces act as a foundation for community bonding and, subsequently, community action. The formation of these digitally-embedded communities has been referred to by anthropologist Aihwa Ong as 'translocal cyberpublics' in global diaspora.

If diaspora peoples are more likely to rely on bonding practices for civic engagement, then the places we gather and build movements are more important than ever: places such as the ESEA Community Centre, Pelican House (a social centre in London for organising and movement building, which serves many communities), Centre 151 (a community centre serving Hackney's Vietnamese, Cambodian and Lao population) and Vietnamese Family Partnership (a London-based Vietnamese community centre), as well as cultural spaces like ESEA Contemporary in Manchester. It's not hard to observe that the majority of the more well-established centres and spaces are found in the UK's capital, although cities with strong ESEA populations, such as Liverpool and Birmingham, have various community centres and services dedicated to specific groups.

The organisation End Violence and Racism Against ESEA Communities (EVR) published a map of ESEA places of interest in the UK in October 2023, which lists over 130 different sites, the majority of which are community centres and religious organisations. However, a recent FOI request showed that public spending on community services and centres dropped by almost a fifth between 2022 and 2024 which, alongside untenable costs for rent, energy bills and staffing, as well as waning donations from individuals who are also feeling the pinch of economic uncertainty, has left many community services extremely vulnerable (Edwards, 2025).

What is clear from the brief overview we have given of some of the forms of organising among ESEA communities in the past few decades is that these movements have always been deeply rooted in workers rights, racial justice and human rights. We should be mindful, therefore, that we do not advocate for certain issues in a vacuum: the global struggle against systems of colonialism is one that is inherently tied to class struggle, racial justice, environmental justice and disability justice. There is much precedent for a united response to human rights issues, using intersectional identities *and* shared interests as a rallying banner. Recent examples of this include the Stand4Uyghur campaign, which has brought Muslims and non-Muslims alike together to protest China's treatment of Uyghur people in Xinjiang, or the mobilisations for Palestinian liberation, which has seen mass support from people of wide-ranging racial or ethnic backgrounds, professions, faiths, gender identities and disabilities.

There are many other organisations and individuals – often unsung heroes – working in activism and advocacy capacities. This work is neither new nor fleeting; it has long existed and will continue in various forms beyond us.

A PROPOSED FRAMEWORK FOR COLLECTIVE AGENCY

To better capture the kind of approach we need, we introduce the following framework. The roles encapsulated within reflect the landscape of ESEA activism as we have experienced it, but the framework itself demonstrates the symbiosis between roles that we envisage.

We call it the Matrix of Collective Agency:

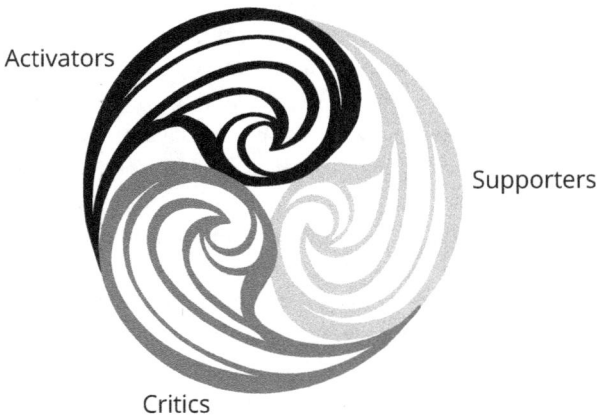

Activators

Supporters

Critics

Figure 8.1 The Matrix of Collective Agency

This illustration draws inspiration from the Mitsudomoe (三つ巴), a traditional Japanese symbol representing the cyclical nature of movement, balance and interdependence. Much like how the Mitsudomoe depicts three swirling forces in dynamic tension, this framework emphasises fluidity in leadership and adaptability in action. By rejecting the illusion of stability, the matrix suggests that agency is sustained through motion, with each role – Activators, Supporters and Critics – equally necessary in the larger flow of collective change. The dream is that one could step back in trust, knowing that others will step forward, reducing the

risk of burnout while reinforcing a culture of shared responsibility. Just as the Mitsudomoe symbol has historically represented resilience and strength – from samurai culture to
spiritual traditions – we hope this model serves as a powerful emblem of collective agency and transformation in our
movements today.

We will dissect the three roles as follows:

- **Activators – The Firestarters and Builders**
  - Activators are the people who initiate, create and drive
    action. They are the strategists, organisers and front-
    line movers who translate ideas into tangible work,
    whether that's through campaigns, direct action, con-
    tent creation, community organising, or building new
    structures of resistance and care. They sustain momen-
    tum by continually pushing forward, often with high
    levels of commitment, passion and personal risk.
  - **Why They Matter:** Without Activators, movements
    stagnate. They transform ideas into reality and ensure
    that the work doesn't remain theoretical or passive.
    Their labour makes abstract goals actionable.
  - **Risks:** Activators are most prone to burnout, disillusion-
    ment and internal conflicts. The pressure to constantly
    be 'on' can result in fatigue, self-sacrifice and frustration
    with slow progress or lack of immediate change. They
    may also feel disconnected from broader perspectives if
    caught up in an echo chamber or the urgency of action.

- **Supporters – The Connectors and Amplifiers**
  - Supporters are the bridge-builders by consuming con-
    tent, attending events and benefiting from community

exchange. They support the cause by way of engagement, amplifying the work towards a wider audience. Though they may not take direct action, they play a crucial role in ensuring momentum, visibility and longevity by translating ideas into accessible narratives and helping extend networks of care and solidarity.

- o **Why They Matter:** Without Supporters, movements struggle to build critical mass. They ensure that the work resonates beyond its immediate circles and that people stay connected, informed and engaged. Their ability to unify and make activism accessible is crucial for longevity and impact.
- o **Risks:** Supporters often operate in spaces where spectatorship can be mistaken for participation. They may feel guilt for not 'doing enough' or frustration when engagement remains superficial. They also risk becoming overwhelmed by the expectation to hold space for others while needing support themselves.

- **Critics – The Observers and Guides**

  - o Critics analyse and provide insight. They take a step back to interpret patterns, assess strategies and articulate broader frameworks that help movements stay accountable, rooted in history and aligned with long-term goals (much as we are doing in this book!). Whether through writing, research, critical discourse, or mentorship, they offer perspective and reflection that prevent movements from becoming short-sighted or reactive.
  - o **Why They Matter:** Without Critics, movements risk repeating past mistakes, losing direction, or burning out without long-term strategy. Their work provides the intellectual, historical and ethical grounding that movements need to sustain themselves and evolve.

o **Risks:** Critics can sometimes become detached from the urgency of action, trapped in analysis without application. If critique is not met with care and accountability, it can discourage rather than support those on the ground. They also risk being dismissed as 'theorists' if their contributions are not well-integrated into movement work.

We should point out that this is not a static, prescriptive framework. Many people shift between roles, and external factors such as economic status, disability and social access can impact one's ability to participate in activism. The goal is not to prescribe fixed identities but to advocate for an approach where everyone can contribute what they can and receive what they need in a way that is not mutually exclusive.

What we envisage as necessary is not for everyone to embody all three roles simultaneously, but to normalise fluidity and exchange between them. We believe that greater knowledge exchange is required between all practices, so we can contribute to a common solution from a space of empathy and mutual knowledge. Furthermore, all actors would benefit from some of the creativity and energy of the other groups, as well as gaining wider perspective and understanding of the challenges they encounter, including burnout, that take a mental and physical toll.

We have repeated throughout this book the reminder that we cannot do everything all at once. But this does not absolve us of our responsibility to stand in solidarity with all forms of resistance to oppression. Indeed, while we are limited by time, by the demands of the state and the economic systems we live in, when we can speak or act, we must ensure that those actions carry intention for *all* struggles under oppression.

How do we do this without burning out?

We might imagine new ways of load sharing that push back against capitalist concepts of productivity, which are driven by our dedication to working weeks of more than 40 hours, overtime and side hustles. This could look like job shares, or rotating roles on a regular basis to ensure time for rest and longevity. We could turn to dedicated, care-focused methods of sharing information and feedback, and expressing dissatisfaction in a regular, nuanced capacity that can address the problems that arise when people's needs aren't met long before breaking point – a capacity that is not afforded by a rapidly moving, impermanent and unstable social grind.

Writer and founder of Peaks of Colour, a Peak District-based nature-for-healing grassroots community group by and for the Global Majority, Evie Muir advocates for a seasonal approach to organising, 'recognising that to survive the seasons you must change with them' (Muir, 2024). Peaks of Colour's core outputs, services and activities are organised during the summer months, and the winter 'hibernation' period is used for foundational, administrative work that cannot be achieved during periods of high activity – with the hope being that the winter hibernation can eventually be used for complete rest and rejuvenation.

Perhaps our vision of greater organising symbiosis is aligned with something akin to Deepa Iyer's Social Changes Ecosystem Map, which outlines a series of roles that allow every person in society to envisage a place for themselves within a movement. Included in the roles are: guides, weavers, experimenters, frontline responders, visionaries, builders, care-givers, disrupters, healers and storytellers (Building Movement Network, 2017).

What we are interested in is how we encourage cohesion and information-sharing between roles in ways that are democratic and rooted in collective care, rather than a culture of individualism or top-down hierarchies. We can draw on the principles of the Social and Solidarity Economy (SSE), a proposed way of organising society that prioritises people and environment over profit. The SSE proposes both systems of workplace management and coordinated revolution that are rooted in the idea of workers councils and popular assemblies. Community councils or neighbourhood assemblies offer an autonomous way for particular workplaces, organising spaces or local areas to be governed democratically by all the people who live and work there, deciding what to create or produce, strategies for action and division of labour. These councils would rely on regular assemblies and rotating, recallable delegates to facilitate communication with other councils on a nationwide scale. These methods of organising are not new, and have cropped up repeatedly throughout history, with present day examples continuing in Chiapas, in southern Mexico, and Rojava, the Democratic Autonomous Administration of North and East Syria.

What is important to remember is that simply raising awareness of things, or modelling them and hoping for a ripple effect, is not going to cause radical, wide-scale adoption. Systems won't just naturally change on their own, especially if there's a bottom line to protect that involves capital. However, we've also talked about the importance of neutralising the harms caused by oppressive systems. It *is* crucial that we find ways of working that don't incapacitate us, cause excessive friction or damage our mental wellbeing: we just shouldn't forget what the overall goal is. It's also really important to *practise* new ways of doing and being, and hone our behavioural shifts, so that we have systems and practices ready to implement when the system does topple. History has taught us that an ill-prepared revolution is bound to fail.

WHEN CHANGE MEETS RESISTANCE

We have spoken throughout this book of various sources of division within ESEA migrant communities of various generations, including the imagined 'we' often used to assume a unifying experience that simply does not exist. Such division and dissonance is also rife in activism spaces. The irony of the severity that can exist within liberation movements that are generally intended to fight for social change and freedom for all, and which often advocate for collective care, makes the burden only heavier. Longtime organisers like adrienne maree brown (2017) have spoken at length about the extent to which we are willing to tear each other down, reiterating that 'uprisings and resistance and mass movement require a tolerance of messiness, a tolerance of many, many paths being walked on at once.'

brown speaks out against the risks of an obsession with critique, admitting:

> We have a way of doing things that is so steeped in critique that I have often wondered if we would strangle movement before it could blossom... Sometimes I think we need to liberate ourselves from critique, both internal and external, to truly give change a chance. (brown, 2017)

If we advocate for collective care, as many in organising movements for social change do, then we must channel it *always*, and not merely when it suits us. We have often wondered, as a collective, in our various messaging groups and informal meets, whether the policing or critiquing behaviour that we witness mainly online (and indeed, in which some of us have also participated in the past and continue to do), is a trauma response to the burden of collective organising. Reactivity, it seems, is often born out of feelings of helplessness – and when we combine this with the online arenas that dominate our lives, it becomes easy to see how policing both the language of others, and the shred of online visibility we are afforded of their actions, allows the enforcers to feel they're doing something helpful.

Perhaps hyper-reactivity is also the natural emotional response during periods of acute crisis. A common refrain we see online is 'the ESEA community has been silent on X cause' or 'ESEA people aren't doing enough for Y movement.' The idea that any individual can be responsible for 'the ESEA community' moving and acting as one is highly unrealistic. In her 2025 book, *Minority Rule*, Ash Sarkar argues that the logic of holding people responsible for what goes on in 'their' community is highly problematic, indicating that:

> [We're] all envoys for a particular tribe... And this simply isn't how community works. We're dispersed by geography, language, class, religion and political values. But instead of imagining anti-racism as something that brings dispersed individuals and communities into contact with one another, it's presented as work that's done in discrete and hermetically sealed groupings. (Sarkar, 2025)

Sarkar goes on to caution against a hyper-reactive, over-policing of others' behaviour online, stating: 'Projection is the dark twin of empathy – rather than extending our own humanity to something outside of ourselves, we force others into becoming totems of our pain' (Sarkar, 2025).

Much anti-racism work today suffers from an over-focus on shared oppression and overlapping experiences rather than on grounding political ideologies from which to organise. Intersectionality simply offers us a lens to understand each other better – it is not the definitive tool we use to dismantle racial capitalism and imperialism. To be clear, we don't shy away from the impact that racism has on an individual's sense of self or ability to assert their identity, and we do believe in the power of mobilisation through collective processing and bonding. But we must not limit ourselves with an inward-looking approach, nor a preoccupation with absolutes, which is heavily present in 'cancel culture', the social phenomenon of en masse boycott, ostracisation or shunning of those who are perceived to have behaved in an unacceptable manner. Both brown and Sarkar question this cycle of blame and shame and whether it amounts to true accountability, or whether it simply creates a situation in which the mass, which envisions itself as hunted prey, behaves like a predator in a 'feeding frenzy' (brown, 2020) or 'an atmosphere of frightened conformity' (Sarkar, 2025).

Many organising efforts we have both witnessed and been part of in the last few years, both online and offline, have suffered, ultimately, from a lack of focus. Those efforts either: a) zero in on the symptom rather than the cause; or b) manifest as surface-level expressions of change and fail to recognise the larger systemic issues at play; or c) operate selectively, choosing only that which is shiny, happy and palatable. Regarding point b), it's important to point out that a tendency to focus on the fun stuff, without interrogation of the struggles and serious issues that underpin them, is often due to the simple fact that organising efforts are overwhelmingly led by people with greater social and economic capital. This makes it difficult to know much about the issues impacting the most marginalised groups in our society.

While finding joy in community and developing a practice rooted in care and wellbeing are critical parts of any movement, a failure to acknowledge deeper systemic harms leaves organisations imbalanced and unimpactful. Sarkar points to this through the words of Barbara Smith, a founding member of the Combahee River Collective, who has suggested that identity politics have become 'a largely middle-class preoccupation' (Sarkar, 2025) in which they are detached from their radical roots. People embrace identity, but conveniently leave out the politics part.

We can illustrate problem a) with examples from our own organising history. In 2020, various organisations, online campaigns and individuals centred around the idea of hate as a virus, against the backdrop of global anti-Asian racism. While it was well intentioned, and certainly made good material for a viral, globally shareable hashtag that originated in the United States (#StopAsianHate), it soon became clear that the positioning of this movement was ill-placed within a longer term context of social change. Before formally confirming our engagements together, various members of besea.n were campaigning together using the hashtag #WeAreNotAVirus. This was an important pushback against a narrative that has long viewed Asian migrants alongside narratives of disease and infection, from Ellis Island detention and disease control measures in New York in the 19th century, to foot-and-mouth scapegoating in the UK in 2001, and global stigma during the 2002–2004 SARS outbreak.

However, as we moved in political and institutional spaces such as various government departments and public institutions, discussing solutions for protecting at-risk, racialised communities, it became quite clear to us that the positioning of 'hate' as a virus was allowing these institutions to sidestep the very apparent existence of actual *racism* as a systemic issue. A virus can be stamped out, whereas racism is insidious, embedded and ungraspable – a much harder pill for the establishment to swallow, let alone admit that institutions under racial capitalism *rely* on racism to function. In the case of #StopAsianHate, it was more than a case of semantics: a focus on COVID-19 racism allowed the institutions supposedly tasked with the care of civilians to completely ignore the needs of racialised communities who were disproportionately impacted by the pandemic once the peak of the crisis was over (even though COVID-19 hasn't actually gone anywhere, much as we like to deny it).

As for problem b), there's an awful lot going on to distract us. In his study of mass protest movements from Brazil to Tunisia in the 2010s, many of which relied on concepts of anti-neoliberal politics, horizontalism and autocracy, Vincent Bevin concluded that these movements failed to achieve true systemic change for the principal reason that most of them were focused on the immediate fight – a reaction to a catalysing event – rather than on organising principles of societal and economic structure for eventual system change. Mass movements that don't have a very clear goal or guiding principles are left liable to have their significance 'imposed upon them from the outside', leaving organisers retroactively grasping for structure (Bevins, 2023).

Alongside declining living standards in the UK and the rise of populist politics, it's clear that, more than ever, mistrust of the state and its institutions, such as the justice system and the police departments, is growing on either side of the political spectrum. For some of us in our own organisation, working within the current system and its promises of 'inclusion' have led to disillusionment: inclusion in what, exactly? However, small, grassroots organisations in our position often buckle under the pressure to continue operating within the confines of our current economic system and what US-based radical feminists of colour organisation INCITE! refers to as the 'non-profit industrial complex'. Within this system, non-profit organisations must operate and are influenced by government policies, corporate interests and capitalist structures, often limiting their ability to create radical change.

To stay afloat, most small organisations must make huge concessions: accept money from restrictive funding pots provided by institutions they mistrust or private companies whose values conflict with their mission, or rely solely on individual donors, which, during a period of economic downturn, has led to many organisations ceasing their operations. What this shows us is that the current funding structures most of us rely on are unsustainable. This operational and financial challenge comes alongside ethical concessions for services (reliance on unethical social media and web platforms, software solutions and finance infrastructure), a pressure to participate in 'professional' practices, and the high risk for burnout, which we have already mentioned. Evie Muir summarises this complex bind perfectly in her case study of the charity or non-profit industrial complex:

> … just as the military industrial complex profits [off] war and the prison industrial complex benefits from crime, the charity industrial complex sentences organisations, no matter how well-intentioned, or how in opposition to the state they wish to be, to the business of *needing* oppression, or being complicit in the state's ongoing oppression, rather than seeking to abolish it. (Muir, 2024)

In their book, *Race to the Bottom: Reclaiming Antiracism*, Azfar Shafi and Ilyas Nagdee highlight the ways in which the charity or NGO industrial complex has been used, both in the racial justice liberation movements of the 1980s and today, as a means to:

> …[buy] activists off the streets and into advocacy roles, tying them down to the structures of funding conditions and charity laws, and reproducing a division of organising work that is shaped more by the dictates of corporate management than movement development. And, by making them accountable to funders and patrons, over the communities they seek to service, these organisations end up forming a layer of professional activists sealed off from a popular base that prevents the formation of radical organisations. (Shafi & Nagdee, 2022)

In order to preserve longevity and stay mission-focused, it's clear to us that those organising in ESEA solidarity and community work today must take care not to succumb to the pitfalls of the new digital age we find ourselves in. While organising within or against the charity or non-profit industrial complex remains fraught, and each organisation or movement will need its own strategy to navigate, the general political shift to the right in the Global North and increasing destabilisation of society means that actors in this space should look to organisations and missions that have endured the decades. These are likely to be small, local community organisations that have not wavered in their mission, no matter the government in place or the shifts in digital trends.

Groups such as Hackney Chinese Community Centre Luncheon Club reforming as ESEA Community Centre Luncheon Club show us that adaptation and change is possible while remaining focused on a globally-informed, local mission. The children and grandchildren of ESEA migrants in the UK have much to learn from elders, particularly as ESEA people; while we can use the language and concepts of the present day to unpick some of the trauma and learned generational behaviours that cause harm, we can also learn more about collectivism and community from those who may have been raised in societies where individualism was not a primary guiding factor in life. Those of us born after the Cold War era (Bevins refers to this period as the 'Americanisation' of the world) have been raised in systems that teach us to view everything as a business opportunity, that everything we do or create must be optimised for maximum production and capital value, 'shaping human beings who think of themselves as autonomous individual firms whose success must be prioritized [sic] above all else – maximizing, optimizing, [sic] hustling, and striving rather than existing as part of any community' (Bevins, 2023).

These are hard shackles to break, but try we must. And finally, we cannot forget the lessons of global history in the struggle against capitalist imperialism. The Black Panther Party is widely seen as one of the strongest connections between the commitment to liberation[4] for Black people and the opposition of western, mainly American, imperialism in the Global South. The BPP 'saw similarities between police brutality against blacks in American ghettos, and the occupation of Vietnam by the American military – and … this empathetic identification was reciprocated by the Vietnamese' (Sarkar, 2025). While it is hard to imagine any kind of anti-imperialist solidarity from Global North governments and public institutions these days, inter-community solidarity is alive and thriving at grassroots level, particularly in the face of rising right wing populism and increasingly restrictive states. We see this in the interconnected, global calls to action against imperialist agendas and legacies in places like Palestine, Myanmar, the Philippines, South Korea, US overseas territories such as Guam, and the Democratic Republic of the Congo.

SOLUTIONS: TOWARDS A RESTORATIVE, TRANSFORMATIVE PRACTICE

Though the landscape may seem bleak and the odds are unfairly stacked, it seems to us as though the criticism often levied against activists is likely to be our salvation: the ability to dream. If our work must endure across decades, how can we ground our work – as advocated for by anti-racism practitioner, Layla F. Saad (2020) – in a practice of good ancestry?

We have no definitive answer, but we have some ideas of where we can look for inspiration. Abolitionist organisers and activists such as Mariame Kaba and Ruth Wilson Gilmore have

---

[4]The term 'liberation' is favoured here, since it indicates a breakaway from a broken system, rather than 'equality' under the existing system.

spent their lives addressing the epidemic of violence under racial capitalism through a practice of transformative and restorative justice, which imagines a world in which carceral systems such as prisons and detention centres do not exist. A key component of this practice is a focus on accountability rather than punishment: abolition encourages us, in the first instance, to undo carceral cultures by identifying punitive logics in our everyday interactions and spaces, and to move away from these punitive responses to harm.

These logics encourage us to root the problem in an individual or a group made up of individuals, and punish them accordingly through a normalisation of isolation and exclusion. In organising spaces, this could look like public shaming online after a perceived failure to say or do a particular thing, denying access to resources or retracting work, or a blanket refusal to work with an organisation based on the action of one of its contributors. Instead, transformative justice practices promote accountability, in which those who cause harm, those who are harmed and those who witness harm address and understand the root causes and the impact of harm, facilitating dialogues in order to move towards reparation.

Though this is easier said than done, hope is being offered through groups like Resist + Renew, who propose grassroots training and facilitation in these contexts, though of course, the nature of volunteering within social advocacy work requires deep and continual mediation between its participants. It's not just a case of scheduling a training session and checking it off the list. Accountability should give individuals the space to learn, and grow a capacity to help dismantle oppressive systems. In our group, we have both participated in punitive logics, and witnessed them from the sidelines, with a common result often being that an individual or organisation retreats completely from any kind of practice, to which we must ask: was it worth it?

While most of us will struggle to shake the culturally ingrained logic of punishment and exclusion, the reflex becomes easier to counter as we witness and experience ways in which individuals and organisations can build systems of safety and care culturally, institutionally and personally. For many of us at besea.n, we have seen firsthand through our work on active bystander training how rooting a practice in care can strengthen feelings of psychological safety for individuals that are much stronger than those cultivated through police presence or reporting systems, for example.

Another key lesson from abolition is the idea of non-reformist reforms: changes that move us closer to dismantling harmful systems rather than reinforcing them. In grassroots organising, especially within existing institutions and the NGO industrial complex, this concept helps us navigate shifts in strategy and approach. While reformist reforms work to maintain the status quo by making oppressive systems more palatable or efficient, non-reformist reforms create meaningful change that weakens those systems and builds alternatives. Various abolitionist scholars, advocates and practitioners have proposed criteria to evaluate whether a reform truly challenges oppressive structures or merely reinforces them. Lawyer and writer Dean Spade has identified the following list for his practice:

Does it provide material relief? Does it leave out an especially marginalized [sic] part of the affected group (e.g., people with criminal records, people without immigration status)? Does it legitimize or expand a system we are trying to dismantle? Does it mobilize people, especially those most directly impacted, for ongoing struggle? (Spade, 2020)

For organisers often moving in crisis mode, where political and social events change on a rapid scale, developing a criteria list for strategy change and decision-making is an excellent line of defence. Regardless of whether their practice is rooted in abolitionist principles or not, social justice practitioners can take lessons from the logic contained within the abolitionist framework. We need long-term, continuous processes made up of smaller steps, rather than short-term, one-off solutions, that recognise the need to act within certain structures, as well as continuing to imagine a future without them.

It is clear that we must move away from polarising approaches and exclusionary practices, and focus on bringing more people into the fold, in recognition that there is a role for everyone in the new world we want to build. Alternative practices may emerge in diverse ways, reflecting the specific challenges and aspirations of different communities. Writer and former worker in the humanitarian crisis sector, Tiara Sahar, shared with us her belief that organising movements must abandon moralising language in order to be able to tip the scale and garner support from people across British society, regardless of their political affiliations.

For example, migrants rights movements advocating for asylum seekers, she argues, often lean heavily on jargon that is either inaccessible to the majority of the public, or acts as a signal of an exclusionary and even elitist way of organising (T. Sahar, personal communication, 7 August 2023). By contrast, polling that uses straightforward language has shown that the majority of people in the UK are sympathetic towards the idea of offering asylum to those who need it. YouGov polling from 2024 asked respondents the question 'Should Britain allow people fleeing persecution or war in other countries to come and live in Britain?', with no use of the word 'asylum'. 55% of people responded that the government should allow either the then existing numbers of people, or more than the existing number. Only 11% of respondents said that none of these people should be granted entry at all (YouGov, 2024).

Since engagement with strategies of resistance to systems of oppression often starts with introspective action, we can return to this practice as a way of dealing with the immediate impulse of reactive behaviour in the polarising, online spaces of today. When we feel anxious about having taken the 'wrong' action or aligned with the 'wrong' viewpoint, it can be helpful to step back from a reactive situation. Instead of getting caught up in the moment, we can immerse ourselves in the knowledge and teachings of those who have been doing this work for a long time. Engaging with these perspectives – whether through writing, audio, or interactive formats – can provide grounding and personal healing. This, in turn, allows us to better navigate the challenges we face every day. At times, we must remember that individual needs and community needs may often be in conflict with each other, and part of the challenge we face is knowing when to remove ourselves from the community space, not only for our own wellbeing and healing, but for the wellbeing of the community itself.

LIBERATION THROUGH CREATIVE PRACTICE

To build the world we want in ways that help ideas, community cohesion and radical acceptance flourish, we can look to creative practices. Creative practices offer us the tools we need for greater understanding, accountability, and making space for connection that spans multiple intersecting identities. In preparation for writing this book, we spoke to several ESEA people whose work lies at the intersection of creative arts and community building. Creative producers and practitioners, Faizal Abdullah and Khai Ramli, who moved to the UK from Singapore, spoke to us about the UK creative scene's potential to challenge assumptions of shared opinions, experiences and identities.

As a Muslim Malay Singaporean in the UK, Khai explained that she does not necessarily feel represented or included in conversations about identity among ESEA groups here. 'I still don't know how to navigate my way, but I will always have these conversations, because I think it's the only way to go', she admitted, adding that she did feel spaces like New Earth and arts groups like Kakilang were open to discussions about these feelings. While Faizal has used his creative practice to explore important topics back home, such as the erosion of Malay language and culture in Singapore in his show '*Siapa Yang Bawa Melayu Aku Pergi? (Who Took My Malay Away?)*', Khai has used event organising through her initiative, RUMAH, to redefine community building among Asian diaspora creatives here in the UK (N. K. Ramli & F. Abdullah, personal communication, 15 June 2023).

Another organisation working across identities and communities is Bitten Peach, the UK's first queer, pan-Asian cabaret company. Through the art of cabaret, drag, burlesque, comedy and spoken word, Bitten Peach seeks not only to provide spaces of queer Asian joy that resist realities of discrimination and marginalisation, but to attempt intercommunity, cross-cultural solidarity among different Asian diaspora through creative performing arts. Member and performer Mahatma Khandi (M. Khandi, personal communication, June 2023) explained to us, 'When I first started doing drag, I was an island, thinking of what I was doing as a way to create space for myself... but as I got older, I was like, wait a minute, I'm not leaving anyone behind.' The idea of 'no one left behind' allows members and performers to develop their own practice as individuals, safe in the knowledge that they are doing so through collective experience and performance in ways that are interchangeable and cyclical.

For Mahatma, engaging with drag performance is not limited to the identities imposed by their Filipino and Sri Lankan heritage ('I'm just a brown clown!'), but offers endless possibilities for connection:

> One of the things that I love the most about [Bitten] Peach is not even the shows. It's the group gatherings, it's the rehearsals, because that's when we get to know each other better and realise the similarities we have, even though we might be all different, different genders, different sexualities, different types of Asians, but at the end of the day, we all eat rice, so that's the same [laughs].

Meanwhile, writer and actor Tuyền Dỗ reflected on the power of storytelling in exploring identity and healing through the arts (T. Dỗ, personal communication, 31 July 2023). She shared with us how her play *Summer Rolls* evolved from what she initially believed would be a lighthearted story to a deeper exploration of family history and intergenerational experiences. Extensive research has shown the benefits of creative practice for healing from different types of trauma, as a mechanism for resisting burnout, and for the development of care-focused community building strategies. Activism has to start from agency, and creative practices provide a space for freedom – to behave, express and exist beyond the constraints of societal norms.

Movement educator and co-author David emphasises that body-centred practices provide an opportunity to rehearse resistance and reimagine reality in ways that transcend intellectual understanding through verbal articulation. Through movement, one can draw from a primal medium of expression to accommodate complexity without the pressures of labels. Regardless of any performative output, the generative nature of creative practices in itself helps with reclaiming power and trust that we all have something to offer. Reinforcing the belief that we all can make a difference, however small, can be the first step away from feelings of helplessness and apathy, the greatest barriers to collective agency and activism.

LIBERATION THROUGH NATURE

Much as we can harness a creative practice to heal from and strategise against the crushing weight of multiple global atrocities and societal decline, so too can we look to another structure that has been in place long before capitalism: the natural world.

Through her work, Potawatomi botanist, author and educator Robin Wall Kimmerer explores the healing power of nature and the idea of liberation through reciprocal relationships with the natural world. Her teachings, rooted in both Indigenous knowledge and scientific ecology, emphasise that nature is not just a resource but a teacher, healer and partner in our collective wellbeing. Her work, rooted in engagement with plants and Indigenous practices, offers reflections on the idea that true liberation comes from rejecting exploitative relationships with the Earth, upon which capitalism is fundamentally reliant, and instead embracing Indigenous ways of knowing, which emphasise respect, reciprocity and kinship.

We see plenty of examples of these kinds of relationships in action. The group Peaks of Colour advocates an alternative route to healing and justice for those impacted by racial and gendered trauma through a series of outdoor 'walkshops', in which Black feminist and abolitionist political theory intersects with embodied connection, community care and nature-allied practices. Along with other initiatives offered by the group, Peaks of Colour believes that this approach offers a challenge to the capitalist, consumerist and colonialist ways in which we are typically told we should interact with nature.

As part of its offering, ESEA Sisters, a volunteer-led collective of women, trans, non-binary and genderqueer ESEA folks, organises nature walks, which combine communal experiences of nature for people of shared ESEA heritage with educational practice, reflection and knowledge exchange in solidarity with different struggles. As we see examples of practitioners and community members returning to nature to ground themselves and their work, we must remain conscious that this, too, is a world under threat: environmental justice intersects with the struggles against racial capitalism, colonialism, gendered violence, economic and food insecurity.

THE FUTURE

The process of writing this book has been many things: revelatory, tumultuous, traumatic, energising and developmental. In many ways, it has been a mirror to the shifts our small group has lived through over the last few years. While we – the co-authors – can speak only to our own experiences, we know that each of the other members and former members will have navigated their own emotional and physical journey that has left its traces on how we see the world. Though largely, we have similar aims and values, we have had to reconcile differences in attitudes, opinions and approaches, particularly when it comes to social and political issues.

Navigating differences against the backdrop of burnout is exhausting. Evie Muir encourages activists to embrace and embody their emotions in their practice, in order to achieve what she calls a restful revolution: exhaustion, grief, rage, anxiety, abundance, joy, hope. Through the work of Buddhist teacher and writer, Lama Rod Owens, Muir explains that, when we attend to these feelings in an embodied, open way, in symbiosis with radical love for ourselves and our communities, emotions like grief and anger can transform into wisdom. 'Capitalism doesn't know how to accommodate such raw and violent emotion,' she says (Muir, 2024).

Something we've always held fast to at besea.n is an approach that is collective rather than hierarchical, with no strict demands on time or commitment. Using this approach has had its many challenges, as well as positives, in a capitalist system that demands productivity and success. We've been down many different avenues: government campaigning, media content, resources, outreach, speaking events and education, facilitation, training, community events, fund sharing and sponsorship, to name a few. Some initiatives have burned us, while others have nourished us.

We've had struggles with funding and tested different modes of organising, trying to find ways that don't rely on operating as a 'business', unwittingly toeing the line of respectability politics in the reformist world of 'diversity, equity and inclusion'. As individuals who have been born into and navigated a world built on capitalist, specifically neoliberal, frameworks, our battle is to prevent ourselves perpetuating the extractive values of this system in our work and in our interactions with each other. It is our hope that we wrote this book, rife with its ups and downs, with a collectivist approach that moves us away from the ideas of ownership and who gets to own, profit or benefit, which can be applied to the work we put out into the world in the future. While we cannot and should not monitor or control the ways in which individuals or groups carry out heritage interpretation, we hope that projects like ESEA Heritage Month will encourage a wide range of exploration on how all ESEA people can use connections over heritage to reach across borders (imagined, physical, symbolic) and build meaningful networks.

Already, we see so much work being carried out by different individuals and collectives across multiple axes, which is truly humbling and beautiful to see. Although much of this chapter has been dedicated to exploring the challenges and harsh realities of activism, particularly in online spaces, we do still believe that the future of online activism can have a place alongside other forms of meaningful action and community building. Virtual organising spaces are occupied by individuals who connect with each other across vast stretches of land, often based on intersecting identities and experiences, in ways not possible before social media.

As far as calls to action go? It would be naive of us not to acknowledge that asking for a lifetime commitment to liberatory practices is a lot. But the vast majority of us can commit to *something* – and something is better than nothing. Most people find that, even though coming to terms with the sheer extent and damage inflicted by systems of harm, it is the work they carry out in community that provides a salve to their open wounds. We recall the words of Angela Davis, who declared, in *Freedom is a Constant Struggle*: 'It is in collectivities that we find reservoirs of hope and optimism' (Davis, 2016).

While we should and must take actions to keep our communities safe, however we define them, we would do well to remember that some communities have different needs, and security, safety and wellbeing may look different to them. By the very nature of the complex means by which they have come into existence, our communities defy reduction into simpler terms, with easy solutions or a 'one size fits all' approach. As we move into new ways of strategising, we remind our readers that activism or grassroots organising aren't simply one type of person, working in a particular way, limited to a certain field.

Many different routes of action are necessary in order to create meaningful change, and each person will thrive in different roles and have different things to contribute. We encourage constant reflections on how, when and where we gather, and the roles that preparing food and breaking bread, creative practice, storytelling, mutual joy and connection with nature have in fostering radical community care and love. These are systems that can sustain us as we begin to reject systems of harm. A collective liberation from oppression will require us to embrace, create with and care for all kinds of people, and so we must make space for the varieties of roles we can all play, even if that looks wildly different to what we imagine for ourselves.

In truth, this book is being written during a period of complete upheaval for besea.n. At the time of writing, we are in full strategic transition, questioning the structures we have been obliged to participate within, and grappling with loss, grief, departures and burnout. To say that writing this book has led to some serious anxieties and feelings of imposter syndrome would be an understatement. Our work has always aimed to embrace what it means to be ESEA in the UK today, and, as we have seen throughout this book, the multiplicity of those experiences have led us down many different paths. As we write, we can quite honestly say that we have no idea what is coming next (and indeed, by publication, some things will already have changed).

Whatever the outcome, we remain hopeful that our approach will be one that is mindful of collective wellbeing, intentional in its focus, and which redefines the very principles of work that have been thrust upon us for as long as we can remember. As ever, we can look to the wisdom of those who came before us for help. adrienne maree brown reminds us: 'We need each other. I love the idea of shifting from "mile wide inch deep movements" to inch wide mile deep movements that schism the existing paradigm' (brown, 2017).

adrienne maree brown's work is deeply informed by the activist and writer Grace Lee Boggs, perhaps one of the most powerful advocates of inter-community solidarity in American civil rights history. Lee Boggs recognised the importance of organising with those around us in her insistence on 'critical connections', that vital practice of base building with others (see Chapter 2: A purpose beyond a label). It is on this hopeful note of connection that we conclude, quite simply, we cannot move in isolation. To progress, we must recognise the invisible thread connecting us to each other and to the Earth – our actions rippling outward, shaping one another.

We have so much more capacity to love each other than we think.

REFERENCES

Adams, T. (2022, 3 March). K-everything: The rise and rise of Korean culture. *The Guardian.* www.theguardian.com/world/2022/sep/04/korea-culture-k-pop-music-film-tv-hallyu-v-and-a

Agarwal, P. (2020). *Wish We Knew What to Say.* Hachette UK.

Age UK. (2020, 30 November). *Ethnic disparities and inequality in the UK: Call for evidence.* www.ageuk.org.uk/siteassets/documents/reports-and-publications/consultation-responses-and-submissions/equality-and-human-rights/age-uks-response-to-ethnic-disparities-and-inequality-in-the-uk-call-for-evidence---nov-2020.pdf

Akala. (2018). *Natives: Race and Class in the Ruins of Empire.* Two Roads.

Albiez-Wieck, S. (2025). Racializing Mestizos and Mestizas in the Philippines—Dean Worcester's Anthropometric Types in the Early 20th Century. *Histories, 5*(2), 23. https://doi.org/10.3390/histories5020023

Amnesty International UK. (2018). *Trapped in the Matrix: Secrecy, Stigma, and Bias in the Met's Gangs Database.* www.amnesty.org.uk/files/reports/Trapped%20in%20the%20Matrix%20Amnesty%20report.pdf

Amnesty International. (2021, 25 May). *Justice for George Floyd: A Year of Global Activism for Black Lives and Against Police Violence.* www.amnesty.org/en/latest/campaigns/2021/05/justice-for-george-floyd-a-year-of-global-activism-for-black-lives-and-against-police-violence/

Anonymous. (2023, June). *Interview with Anonymous* (Amy, Interviewer) [Personal communication].

Asia Media Centre. (2019, 23 May). A guide to using Asian names. *Asia Media Centre, New Zealand.* www.asiamediacentre.org.nz/features/a-guide-to-using-asian-names/

Asia Society. (2017, 25 April). *35 years after Vincent Chin's murder, how has America changed?* https://asiasociety.org/blog/asia/35-years-after-vincent-chins-murder-how-has-america-changed

Asian Century Institute. (2014). *South Korea's plastic surgery obsession*. https://asiancenturyinstitute.com/society/272-south-korea-s-plastic-surgery-obsession

Aujla-Sidhu, G. (2021). *The BBC Asian Network: The Cultural Production of Diversity*. London: Palgrave Macmillan.

Baker, C. (2023). *NHS staff from overseas: Statistics. Commons Library, 7783*. https://commonslibrary.parliament.uk/research-briefings/cbp-7783/

Balčaitė, I. (2019). Networks of resilience: Legal precarity and trans-border citizenship among the Karen from Myanmar in Thailand. *TRaNS: Trans-Regional and -National Studies of Southeast Asia, 7*(1), 63–89. https://doi.org/10.1017/trn.2018.12

Banfield-Nwachi, M. (2022, 26 December). 'Pulled from both sides': Britons who send money abroad face living costs squeeze. *The Guardian*. www.theguardian.com/business/2022/dec/26/pulled-from-both-sides-britons-who-send-money-abroad-face-living-costs-squeeze

Barber, T. (2018). The integration of Vietnamese refugees in London and the UK. *Research Gate*, UNU WIDER. 10.35188/UNU-WIDER/2018/444-5

Barghouti, O. (2011). *BDS: Boycott, Divestment, Sanctions – The Global Struggle for Palestinian Rights*. Haymarket Books.

Barkawi, T. (2006). *Globalization and War*. Rowman & Littlefield.

BBC News. (2022, 11 February). Labour MP Neil Coyle suspended over racist comment claim. *BBC News*. www.bbc.co.uk/news/uk-politics-60347836

BenarNews. (2019, May 21). Southeast Asia struggles with trash rejected by China. *Eco-Business*. www.eco-business.com/news/southeast-asia-struggles-with-trash-rejected-by-china/

Benjamin, R. (2022). *Viral Justice*. Princeton University Press.

Benton, G. (2007). *Chinese Migrants and Internationalism*. Routledge.

Benton, G., & Gomez, E. T. (2007). *The Chinese in Britain, 1800–Present: Economy, Transnationalism, Identity*. Palgrave Macmillan.

Berry, C. (2021, 23 April). *Rethinking class in the age of rent*. Autonomy. https://autonomy.work/portfolio/berry-class-rent/#1592467993162-2945af24-5883

besea.n (2020). *No our food is not dirty*. www.besean.co.uk/spotlight/no-our-food-is-not-dirty-eseaeats#:~:text=MasterChef's%20

casual%20cultural%20insensitivity%20towards%20East%20
and,connotations%20seen%20on%20Chef%20Philli's%20ins-
tagram%20profile

Bevins, V. (2023). *If We Burn*. Public Affairs.

Bhugra, D., & Becker, M. A. (2005). Migration, cultural bereavement
and cultural identity. *World Psychiatry, 4*(1), 18. National Library of
Medicine. https://pmc.ncbi.nlm.nih.gov/articles/PMC1414713/

Blyth, M. (2013). *Austerity: The History of a Dangerous Idea*. Oxford
University Press.

Boggs, G. L., & Kurashige, S. (2012). *The Next American Revolution:
Sustainable Activism For The Twenty-First Century*. University of
California Press.

brown, a. m. (2017). *Emergent Strategy: Shaping Change, Changing
Worlds*. AK Press.

brown, a. m. (2020). *We Will Not Cancel Us: And Other Dreams of
Transformative Justice*. AK Press.

Building Movement Network. (2017). *Social change ecosystem map*.
Building Movement Project. https://buildingmovement.org/our-
work/movement-building/social-change-ecosystem-map/

Burack, E. (2023, 2 February). The royal family showed up for
Buckingham Palace reception celebrating Asian communities. *Town & Country*. www.townandcountrymag.com/society/
tradition/g42744283/royal-family-buckingham-palace-
reception-asian-community-february-2023-photos/

Burgess, S., & Greaves, E. (2013). Test scores, subjective assessment, and stereotyping of ethnic minorities. *Journal of Labor
Economics, 31*(3), 535–576. https://doi.org/10.1086/669340

Butler, J. (1990). *Gender Trouble: Feminism and the Subversion of
Identity*. Routledge.

Cabinet Office. (2021). Writing about ethnicity. *Ethnicity facts and
figures*.   www.ethnicity-facts-figures.service.gov.uk/style-guide/
writing-about-ethnicity

Cambria Daily Leader. (1919, 14 June). Race riots. *The Cambria
Daily Leader*. The National Library of Wales. https://newspapers.
library.wales/view/4427229/4427230/33/

Canary. (2025, 12 March). Allianz and Aviva facing injunctions from
Palestine Action over ties to Israeli weapons manufacturer. *Canary*.

www.thecanary.co/uk/analysis/2025/03/12/allianz-aviva-palestine-action/

Carbon Brief. (2023). *The Carbon Brief profile: China.* https://interactive.carbonbrief.org/the-carbon-brief-profile-china/index.html

Carlile, C., & Harrison, R. (2022). Addressing subtle forms of hate in UK media coverage of migration. *Ethical Consumer.* https://research.ethicalconsumer.org/sites/default/files/inline-files/Addressing-subtle-hate-UK-media-coverage-migration-Full-Report-Oct2022.pdf

Casci, M. (2006, 26 September). Seven held as police target cannabis factories. *Bradford Telegraph and Argus.* www.thetelegraphandargus.co.uk/news/938202.seven-held-as-police-target-cannabis-factories/

Centre for Economic Policy Research (CEPR). (2024). *China is the world's sole manufacturing superpower.* https://cepr.org/voxeu/columns/china-worlds-sole-manufacturing-superpower-line-sketch-rise

Center for Preventive Action. (2024, 17 September). Territorial disputes in the South China Sea. *Global Conflict Tracker, Council on Foreign Relations.* www.cfr.org/global-conflict-tracker/conflict/territorial-disputes-south-china-sea

Chae, D. H., Yip, T., Martz, C. D., Chung, K., Richeson, J. A., Hajat, A., et al. (2021). Vicarious racism and vigilance during the COVID-19 pandemic: Mental health implications among Asian and Black Americans. *Public Health Reports, 136*(4), 508–517. https://doi.org/10.1177/00333549211018675

Chamberlain, D. (2021, 25 May). *Collections uncovered: Omdurman battlefield skulls.* Anatomical Museum, University of Edinburgh. https://anatomicalmuseum.wordpress.com/2021/05/25/collections-uncovered-omdurman-battlefield-skulls/#_ftn3

Chan, A. (2015). *China's Workers Under Assault : The Exploitation of Labor in a Globalizing Economy.* Routledge.

Chee, H. L. (2015). Medical tourism and the sexualization of the Thai body. *Global Public Health, 10*(2), 211–224.

Chen, T. (2021, 6 January). Three women accused of whitewashing Mahjong say they're sorry for not paying homage to its Chinese origins. *BuzzFeed News*. www.buzzfeednews.com/article/tan yachen/the-mahjong-line-criticism

Cheng, F. (2016). *Constructing a new Asian masculinity: Reading lilting against other films by Asian filmmakers*. https://scholarsbank. uoregon.edu/items/385e6c0e-245e-43b0-99b5-f430de5d7c77

Cheryan, S., & Monin, B. (2005). Where are you *really* from?: Asian Americans and identity denial. *Journal of Personality and Social Psychology*, *89*(5), 717–730. https://doi.org/10.1037/0022-3514.89.5.717

Choy, C. C. (2003). *Empire of Care: Nursing and Migration in Filipino American History*. Duke University Press.

Commission on Race and Ethnic Disparities. (2021). *Commission on Race and Ethnic Disparities: The Report*. https://assets.publishing.service.gov.uk/government/uploads/system/uploads/attachment_data/file/974507/20210331_-_CRED_Report_-_FINAL_-_Web_Accessible.pdf

Committee of 100. (2017, 1 May). Chinese American journeys: Joan Chen, pioneering actress. *Committee of 100*. YouTube. www.youtube.com/watch?v=q5yOH9DDQDU

Crenshaw, K. (1991). Mapping the margins: Intersectionality, identity politics, and violence against women of color. *Stanford Law Review*, *43*(6), 1241–1299. https://doi. org/10.2307/1229039

Curington, C., Lin, K.-H., & Lundquist, J. (2015, 1 July). *Dating partners don't always prefer 'their own kind': Some multiracial daters get bonus points in the dating game*. https:// thesocietypages.org/ccf/2015/07/09/dating-partners-dont-always-prefer-their-own-kind-some-multiracial-daters-get-bonus-points-in-the-dating-game/

Curington, C. V., Lundquist, J. H., & Lin, K.-H. (2021). *The Dating Divide: Race and Desire in the Era of Online Romance*. University of California Press.

daikon* Zine. (2018, 5 September). Queer / Trans Survey. *daikon* Zine*. https://daikon.co.uk/blog/queer-trans-survey

Davis, A. (2016). *Freedom is a Constant Struggle: Ferguson, Palestine, and the Foundations of a Movement*. Haymarket Books.

De Pacina, M. (2023, 12 July). Video: TikTok user calls out Asian women's 'internalized racism' against dating Asian men. *NextShark*. https://nextshark.com/tiktok-internalized-racism-dating-asian-men

Diversity UK. (2023, 20 March). *Census 2021 data reveals ethnic make up of UK population.* https://diversityuk.org/census-2021-data-reveals-ethnic-make-up-of-uk-population/#:~:text=The%20largest%20increases%20were%20seen

Dỗ, T. (2023, 31 July). *Interview with Tuyền Dỗ* (I. Pan, Interviewer) [Personal communication].

Dragons and Lions. (2021, 17 June). *About – Dragons and Lions.* https://dragonsandlions.co.uk/about

Dugan, E. (2022, 14 August). Revealed: Indonesian workers on UK farm 'at risk of debt bondage'. *The Guardian*. www.theguardian.com/uk-news/2022/aug/14/uk-farm-workers-kent-debt-indonesian-brokers

Echo, E. (2023, 23 June). *Interview with Eva Echo* (D. Kam, Interviewer) [Personal communication].

ECPAT International. (2016, May). *Global Study on Sexual Exploitation of Children in Travel and Tourism – Full Report.* ECPAT. https://ecpat.org/resource/the-global-study-on-sexual-exploitation-of-children-in-travel-and-tourism/

Edwards, L. (2025, 28 January). Winner of BBC Interior Design Matters helps revamp community space. *The Mirror*. www.mirror.co.uk/news/uk-news/community-centre-revitalised-help-bbc-34562363

El-Enany, N. (2020, 29 April). *Europe's colonial embrace and the Brexit nostalgia for empire are two sides of the same coin.* LSE. https://blogs.lse.ac.uk/brexit/2020/04/29/europes-colonial-embrace-and-brexit-as-nostalgia-for-empire-are-part-of-the-same-story/

Endicott, K. (2016). Malaysia's Original People: Past, Present and Future of the Orang Asli. In *JSTOR*. NUS Press. https://www.jstor.org/stable/j.ctv1qv35n

English Heritage. (n.d.). *The Ayahs' Home.* Blue Plaques. www.english-heritage.org.uk/visit/blue-plaques/ayahs-home

ESEA Music. (2023). '(Re)*Orientated' An ESEA Music Survey*. www. eseamusic.co.uk/reorientated

Espiritu, Y. L. (1999). Gender and labor in Asian immigrant families. *American Behavioral Scientist, 42*(4), 628–647. https://doi. org/10.1177/00027649921954390

Evans, N. (1980). *The South Wales race riots of 1919*. Llafur, *3*(1), 5.

Feminism Fierce. (2021, 26 June). '*The distinction is important!' Instagram*. www.instagram.com/feminismfierce/p/CQmClmBHHPu/?hl=en

Fischer, P. (2015, 22 February). The Korean Republic of New Malden: How Surrey became home to the 70. *The Independent*. www.independent.co.uk/news/uk/home-news/the-korean-republic-of-new-malden-how-surrey-became-home-to-the-70-yearold-conflict-10063055.html

Freeland, L.S. (1951). *Language of the Sierra Miwok*. Waverly Press.

Frias, L. (2021, 6 May). Boba liberalism: Critique on shallow political identity amid AAPI crisis. *Business Insider*. www. businessinsider.com/boba-liberalism-critique-on-a-shallow-political-identity-amid-crisis-2021-3

Goh, J. X., & McCue, J. (2021). Perceived prototypicality of Asian subgroups in the United States and the United Kingdom. *Journal of Experimental Social Psychology, 97*, 104201. https://doi. org/10.1016/j.jesp.2021.104201

Goltz, D. B. (2009). *Queer Temporalities in Gay Male Representation*. Routledge.

Golzar Anderson, M., & Phan, T. (2022, 13 July). Decolonisation on (in) the ground: Farming as colonial resistance in Palestine. *Shado Magazine*. https://shado-mag.com/columns/land-defenders/decolonisation-colonial-resistance-in-palestine/

Gonzalez, C. G. (2021). Racial capitalism and the Anthropocene. *The Cambridge Handbook of Environmental Justice and Sustainable Development*, 72–85. https://doi.org/10.1017/9781108555791.007

Google Trends. (2025). Google trends. https://trends.google. com/trends/explore?q=hongkonger

Gordon, A. (2024, 17 May). King Charles III's net worth has increased dramatically, U.K. Rich List reveals. *TIME*. https://time. com/6979293/king-charles-net-worth-2024/

Gram, L., & Mau, A. (2024, 25 April). 'We are not the virus' – *Experiences of racism among East & Southeast Asian heritage young people in London during the height of the COVID-19 pandemic.* besea.n. https://doi.org/10.1371/journal.pgph.0002016

Greater London Authority. (2021). *Mayor's Diversity Commission to celebrate London's Untold Stories.* www.london.gov.uk/press-releases/mayoral/mayors-commission-to-celebrate-londons-stories

Gregoriou, C., Ras, I. A., & Muždeka, N. (2021). 'Journey into hell [...where] migrants froze to death': A critical stylistic analysis of European newspapers' first response to the 2019 Essex Lorry deaths. *Trends in Organized Crime, 25,* 318–337. https://doi.org/10.1007/s12117-021-09418-x

Guillermo, E. (2014, 25 May). Elliot Rodger's manifesto shows self-hate fueled anti-Asian violence that kicked off Isla Vista rampage. *AALDEF Blog.* www.aaldef.org/blog/elliot-rodgers-manifesto-shows-self-hate-fueled-anti-asian-violence-that-kicked-off-isla-vista-rampa/

Hackney Chinese Community Services Association (2023) *Met Police Investigation Outcomes on Hate Crime towards East and Southeast Asians (ESEA).* www.hackneychinese.org.uk/post/met-police-investigation-outcomes-on-hate-crime-towards-east-and-southeast-asians-esea

Haddad, M., & Hussein, M. (2021, 10 September). Infographic: History of US interventions in the past 70 years. *Al Jazeera.* www.aljazeera.com/news/2021/9/10/infographic-us-military-presence-around-the-world-interactive

Haffety, T. (2021, 22 December). *Neither here nor there.* besea.n. www.besean.co.uk/spotlight/neither-here-nor-there

Halberstam, J. (1998). *Female Masculinity.* Duke University Press.

Halberstam, J. (2005). *In a Queer Time and Place: Transgender Bodies, Subcultural Lives.* New York University Press.

Hall, S. (1973). *Encoding and Decoding in the Television Discourse.* Birmingham University.

Hall, S. (1987). *Minimal selves. In 'Identity the real me'* (ICA document no. 6). Institute of Contemporary Arts.

Hamdache, B. (2022, 30 November). 2021 census: Distorted reporting echoes racist conspiracy. OpenDemocracy. www.opendemocracy.net/en/2021-census-great-replacement-the-ory-racist-data-media-white-christians-england-wales/

Hampton, M. P. and Hamzah, A. (2016) Change, choice, and commercialization: Backpacker routes in Southeast Asia. *Growth and Change*, *47*(4), 556–571. ISSN 0017-4815.

Hancox, D. (2021, 25 May). The secret deportations: How Britain betrayed the Chinese men who served the country in the war. *The Guardian*. www.theguardian.com/news/2021/may/25/chinese-merchant-seamen-liverpool-deportations

Hirsch, A. (2018). *Brit(ish): On Race, Identity and Belonging*. Vintage.

hooks, b. (1994). *Teaching to Transgress: Education as the Practice of Freedom*. New York: Routledge.

House of Commons Library. (2021). The Crown Estate. *House of Commons Library*. https://commonslibrary.parliament.uk/research-briefings/sn02950/

House of Lords Library. (2024). Budget 2024: Impact on the cultural sector. *House of Lords Library*. https://lordslibrary.parliament.uk/budget-2024-impact-on-the-cultural-sector/#:~:text=In%20recent%20years%2C%20the%20public,museums%20increased%20their%20earned%20income

Hui, A. (2020, 23 August). London no longer has one Chinatown. *It has many*. Resy. https://blog.resy.com/2020/08/london-no-longer-has-one-chinatown-it-has-many/

Hui, A. (2023). *Takeaway*. Trapeze.

Human Dignity Trust. (2024). *Map of countries that criminalise LGBT People*. Human Dignity Trust. www.humandignitytrust.org/lgbt-the-law/map-of-criminalisation/?type_filter_submitted=&type_filter%5B%5D=crim_lgbt

Human Rights Watch. (2024, 2 Sept). *After riots, UN calls on UK to tackle systemic racism, colonial legacies*. www.hrw.org/news/2024/09/02/after-riots-un-calls-uk-tackle-systemic-racism-colonial-legacies

Humi, F. (2023, 7 August). *Interview with Francesca Humi* (M.-A. Vu Peterson, Interviewer) [Personal communication].

Hyun, J. (2019, 15 January). White-owned pho restaurant that tried to trademark the word 'pho' under fire for 'vegan pho.' *NextShark*. https://nextshark.com/pho-restaurant-sue-vietnamese

Hyunh, C. and Wemyss, G. (2021). *Invisibility and Hypervisibility: East and Southeast Asian Racism in the UK during COVID-19*. London: Centre for Research on Migration, Refugees and Belonging, University of East London.

Ipsos. (2020, 15 June). *Attitudes to Race and Inequality in Great Britain*. www.ipsos.com/en-uk/attitudes-race-and-inequality-great-britain

Jenn. (2014, 4 April). Hey Air France, your Orientalism is in the air | #FixedIt4UAF @AirFrance. *Reappropriate*. https://reap-propriate.co/2014/04/hey-air-france-your-orientalism-is-in-the-air/

Joint Council for the Welfare of Immigrants (JCWI). (2024, 11 June). *The Hostile Environment explained*. https://jcwi.org.uk/reportsbriefings/the-hostile-environment-explained/

Jolly, K. (2021, 14 January). *WhatsApp interview* (M.-A. Vu Peterson, Interviewer) [Personal communication].

Kelly, A., & McNamara, M. -L. (2015, 23 May). 3,000 children enslaved in Britain after being trafficked from Vietnam. *The Guardian*. www.theguardian.com/global-development/2015/may/23/vietnam-children-trafficking-nail-bar-cannabis

Khandi, M. (2023, June). *Interview with Mahatma Khandi* (D. Kam, Interviewer) [Personal communication].

Khomami, N., & Hawkins, A. (2023, 20 June). Translator alleges work on Chinese radical 'plagiarised' in British Museum show. *The Guardian*. www.theguardian.com/culture/2023/jun/20/translator-alleges-work-on-chinese-radical-plagiarised-in-british-museum-show

kimchiwangmandu. (2023, 17 January). I finally met my KOREAN MILITARY BOYFRIEND *emotional* | NYC to Korea VlogKR ldr meeting first time. YouTube. www.youtube.com/watch?v=Eb7N4EnE29A

King, R. (2013, 20 November). The uncomfortable racial preferences revealed by online dating. *Quartz*. https://qz.com/149342/the-uncomfortable-racial-preferences-revealed-by-online-dating

Kipling, R. (1899). *The White Man's Burden*. Doubleday & Company.

Kosasie, V. (2023, June). (Amy, Interviewer) [Personal communication].

Kozuch, E. (2019, 22 May). *API LGBTQ youth at heightened risk for discrimination*. Human Rights Campaign. www.hrc.org/press-releases/hrc-uconn-survey-finds-api-lgbtq-youth-at-height-ened-risk-for-discriminatio

Kushner, T. (2006) *Remembering Refugees: Then and Now*. Manchester University Press.

Kwan, J. (2023, June). *Interview with Jason Kwan* (Amy, Interviewer) [Personal communication].

Kwek, N. (2024, 25 February). Is Newington Scotland's new Chinatown? We take a look at the influx of Chinese eateries in this Edinburgh neighbourhood. *The Scotsman*. www.scotsman. com/lifestyle/food-and-drink/is-newington-scotlands-new-chi-natown-we-take-a-look-at-the-influx-of-chinese-eateries-in-this-edinburgh-neighbourhood-4531775

Lakhani, N., Gayle, D., & Taylor, M. (2023, 12 October). How criminalisation is being used to silence climate activists across the world. *The Guardian*. www.theguardian.com/environment/2023/oct/12/how-criminalisation-is-being-used-to-silence-climate-ac-tivists-across-the-world

Lam, J. (2014, 19 March). 'Hongkonger' makes it to world stage with place in the Oxford English Dictionary. *South China Morning Post*. https://web.archive.org/web/20190730063432/https://www.scmp.com/news/hong-kong/article/1451929/finally-hongkonger-arrives-world-stage

Lavelle, D. (2024, 23 September). Cost of taxpayer-funded grant for UK monarchy to rise by £45m. *The Guardian*. www.the-guardian.com/uk-news/2024/sep/23/cost-of-grant-that-funds-uk-monarchy-to-rise-by-more-than-53

Lavrakas, P. J. (2008). *Encyclopedia of Survey Research Methods*. Sage.

Layput. (2021, 8 August). *Ivy Asia Chelsea racist ad*. www.youtube. com/watch?v=jDX48-pyYaI

Le, T. P., & Ahn, L. H. (2024). Asian American women's racial dating preferences: An investigation of internalized racism, resistance and empowerment against racism, and desire for status. *Sex Roles, 90.* https://doi.org/10.1007/s11199-024-01450-9

Leary, A., Maxwell, E., Myers, R., & Punshon, G. (2024). Why are healthcare professionals leaving NHS roles? A secondary analysis of routinely collected data. *Human Resources for Health, 22*(1). https://doi.org/10.1186/s12960-024-00951-8

Lee, G. (2021, 26 March). *#chinesevirus: The long racism that lurks behind COVID-19.* Postcolonial Politics. https://postcolonialpolitics.org/chinesevirus-racism-behind-covid-19/

Lee, G. (2023, 6 August). *Narrating and displaying China and Chineseness: White dominance, white saviourism and decoloniality.* Postcolonial Politics. https://postcolonialpolitics.org/narrating-and-displaying-china-and-chineseness-white-dominance-white-saviourism-and-decoloniality/

Lee, J. (2023). *Biting the Hand.* Henry Holt and Company.

Lee, J. J. (2024). *Dispersals.* Penguin.

Liberty. (2021). *Liberty's Briefing on the Police, Crime, Sentencing and Courts Bill for Report Stage in the House of Commons.* www.libertyhumanrights.org.uk/wp-content/uploads/2020/04/Libertys-briefing-on-the-Police-Crime-Sentencing-and-Courts-Bill-Report-Stage-HoC-July-2021.pdf

Light, M. (2020, July). Colonialism, hegemony, and the environment. *Climate Just Collective.* www.climatejustcollective.co.uk/colonialism-hegemony-and-the-environment

Lin, T. (2022). Racial triangulation revisited: The racial positioning of Asian Americans. *Sociology of Race and Ethnicity, 8*(1), 6–22.

Lisney, E. (2023, 19 June). (D. Kam, Interviewer) [Personal communication].

Liverpool John Moore University (LJMU). (2023). *Liverpool's history with the Chinese community.* www.ljmu.ac.uk/about-us/news/articles/2023/1/19/liverpools-history-with-the-chinese-community

Lorde, A. (1984). *The Master's Tools Will Never Dismantle the Master's House.* Penguin Books.

Lorenzo, F. M. E., Galvez-Tan, J., Icamina, K., & Javier, L. (2007). Nurse migration from a source country perspective: Philippine country case study. *Health Services Research, 42*(3p2), 1406–1418. https://doi.org/10.1111/j.1475-6773.2007.00716.x

Louie, S. (2014, 25 June). Elliot Rodger's Asian self-hatred. *Psychology Today*. www.psychologytoday.com/gb/blog/minority-report/201406/elliot-rodgers-asian-self-hatred

Lovell, J. (2014, 30 October). The Yellow Peril: Dr Fu Manchu & the Rise of Chinaphobia by Christopher Frayling – review. *The Guardian*. www.theguardian.com/books/2014/oct/30/yellow-peril-dr-fu-manchu-rise-of-chinaphobia-christopher-frayling-review

Lucas, J. (2015, 17 April). *Rising rents might mean the end of London's Chinatown*. VICE. www.vice.com/en/article/threat-to-londons-chinatown-304/

Luo, M., Qi, S., & Du, W. (2023). Family and queer temporality among Chinese young gay men: Yes, but not yet. *The Sociological Review*, *72*(1), 137–154. https://doi.org/10.1177/00380261231156274

Ma, G. (2022, 18 September). *Chinese Chippy Girl Podcast* (Season 3 Episode 2) [Podcast]. Acast. https://shows.acast.com/chinese-chippy-girl/episodes/6336b8fedf4c79001271aa46

Mac, J., & Smith, M. (2018). *Revolting Prostitutes: the Fight for Sex Workers' Rights*. Verso.

Mark, J. (2021, 22 July). 'Queen of Congee' apologizes for cultural appropriation but still sells the 'improved' Asian dish. *The Washington Post*. www.washingtonpost.com/nation/2021/07/22/congee-queen-karen-taylor/

Marsha, A. (2018, 9 January). *What's the deal with men's rights activists and Asian fetishes?* VICE. www.vice.com/en/article/whats-the-deal-with-mens-rights-activists-and-asian-fetishes/

Mass, A. (1992). Interracial Japanese Americans. In M. Root (Ed.), *Racially Mixed People in America*. Sage.

McCoy, A.W. (2009). *Policing America's Empire: The United States, the Philippines, and the Rise of the Surveillance State*. University of Wisconsin Press.

Menon, A. V. [@alokvmenon]. (2024, 21 March). There is magic here, beneath all the shame. *beauty is our heritage. thank you @ brownkindskin for championing those of us...* [Video]. Instagram. www.instagram.com/p/C4yud4lurA0/

Mercer, D. (2021, 5 February). COVID-19: NHS staff fall victim to anti-Chinese hate crimes – amid fears violence will rise when lockdown ends. *Sky News*. https://news.sky.com/story/covid-19-nhs-staff-fall-victim-to-anti-chinese-hate-crimes-amid-fears-violence-will-rise-when-lockdown-ends-12206686

Mirza, M. (2017, 11 September). Lammy review: The myth of institutional racism. *Spiked.* www.spiked-online.com/2017/09/11/lammy-review-the-myth-of-institutional-racism/

Mishra, S. (2021, 19 March). Calls for Atlanta police captain to resign as 'China virus' posts emerge after massage parlor killings. *The Independent.*www.independent.co.uk/news/world/americas/atlanta-salon-shooting-stop-asian-hate-b1818846.html

Mohanty, C. T. (1984). Under Western eyes: Feminist scholarship and colonial discourses. *boundary 2, 12*(13), 333–358.

Moosavi, L. (2020). 'Can East Asian students think?': Orientalism, critical thinking, and the decolonial project. *Education Sciences, 10*(10), 286. https://doi.org/10.3390/educsci10100286

Morin, N. (2020, 12 May). K-Pop's beautiful men are breaking the rules of masculinity, but can America handle it? *Refinery29.* www.refinery29.com/en-us/2020/05/9674149/kpop-male-singers-masculinity

Morris, N. (2022). *MIXED/OTHER: Explorations of Multiraciality in Modern Britain.* Trapeze.

Muir, E. (2024). *Radical Rest.* Elliot & Thompson.

Nam, M. (2023, 12 June). *Interview with Mark Nam* (D. Kam, Interviewer) [Personal communication].

National Trust. (2024, 1 May). *Make their year.* Facebook. www.facebook.com/nationaltrust/videos/1621431501957289

Ng, K. (2021, 4 January). Attack on London law student over coronavirus was 'racially motivated'. *The Independent.* www.independent.co.uk/news/uk/crime/coronavirus-attack-london-racism-jonathan-mok-b1782233.html

Ng, V.-L. (2023, 7 August). *Untitled* (M.-A. Vu Peterson, Interviewer) [Personal communication].

Ngai, M. M. (2004). *Impossible Subjects: Illegal Aliens and the Making of Modern America.* Princeton University Press.

Nguyen Phan, Q. M. (2023). *Dust Child.* Simon and Schuster.

Nguyen, Y. (2021, 21 March). The long history of sexual and physical violence Asian women face in the U.S. (L. Garcia Navarro,

Interviewer). In *NPR*. www.npr.org/2021/03/21/979683478/the-long-history-of-sexual-and-physical-violence-asian-women-face-in-the-u-s

Nuallak, P. (2022). *Pearls from Their Mouth*. Hajar Press.

Nuallak, P. (2023). *ESEA FOLKS 4 …?* Montez Press. https://montez-press.com/static/media/interjectionPDFS/Interjection-009-11_Pear_Nuallak.pdf

Nuallak, P., & *Remember & Resist*. (2024, 17 July). *Yellow Peril Self-Awareness Manual*. Pear Nuallak. https://pearnuallak.com/yellow-peril-self-awareness-manual/

Nugroho, J. (2016, 29 July). The hidden histories of homosexuality in Asia. *Fair Observer*. www.fairobserver.com/region/asia_pacific/hidden-histories-homosexuality-asia-77120/

Nunag, L. (2020, 31 July). Being a migrant worker, being able to say 'no' at work, and how nursing is exposing inequalities with Lloyd Nunag (O. Gagan, Interviewer). www.oliviagagan.com/emotional-labour-articles/dorcas-boamah-aafks

Nye, J. S. (1990). Soft power. *Foreign Policy*, *80*(80), 153–171.

Office for National Statistics. (ONS). (2022). *Ethnic group, England and Wales: Census 2021*. www.ons.gov.uk/peoplepopulationand community/culturalidentity/ethnicity/bulletins/ethnicgroupeng landandwales/census2021

Office for National Statistics (ONS). (2023, 29 November). *Ethnicity pay gaps, UK: 2012 to 2022*. www.ons.gov.uk/employmentand-labourmarket/peopleinwork/earningsandworkinghours/articles/ethnicitypaygapsingreatbritain/2012to2022

Olusoga, D. (2016) *Black and British: A Forgotten History*. London: Pan Macmillan.

ONS Census Transformation Programme. (2016). Assessment of initial user requirements on content for England and Wales. In *Assessment of Initial User Requirements on Content for England and Wales*. Office of National Statistics.

Oyen, M. (2013). Fighting for equality: Chinese seamen in the Battle of the Atlantic, 1939–1945. *Diplomatic History*, *38*(3), 526–548. https://doi.org/10.1093/dh/dht106

Pai, H. -H. (2008). *Chinese Whispers: The True Story Behind Britain's Hidden Army of Labour*. Penguin.

Park, G. (2020). *Not Your Yellow Fantasy*. New Degree Press.

Park Hong, C. (2020). *Minor Feelings: An Asian American Reckoning*. One World.

Parker, D., & Song, M. (2001). *Rethinking Mixed Race*. Pluto Press.

Parker, D., & Song, M. (2006). Ethnicity, social capital and the internet. *Ethnicities*, *6*(2), 178–202. https://doi.org/10.1177/1468 796806063751

Parker, D., & Song, M. (2007). Inclusion, participation and the emergence of British Chinese websites. *Journal of Ethnic and Migration Studies*, *33*(7), 1043–1061. https://doi.org/10.1080/ 13691830701541564

Parliament. (2020, May). *Written evidence submitted by the Filipino UK Nurses Association* (MRS0431). Written Evidence. https:// committees.parliament.uk/writtenevidence/3805/pdf/

Petersen, W. (1966, 9 January). Success story, Japanese-American style. *The New York Times Magazine*.

Pistol, R. (2017). *Internment During the Second World War: A Comparative Study of Great Britain and the USA*. London: Bloomsbury Academic.

Public Data Lab. (2021, March). *East and Southeast Asians: Documenting a Category in the Making*. Public Data Lab. https:// publicdatalab.org/projects/esea

Ramli, N. K., & Abdullah, F. (2023, 15 June). (D. Kam, Interviewer) [Personal communication].

Reid, S. (2023, 9 August). *Interview with Sarah Reid* (K.-W. Hii, Interviewer) [Personal communication].

Richards, W. (2024, 8 October). The Great Escape festival no longer sponsored by Barclays following boycott. *Rolling Stone UK*. www. rollingstone.co.uk/music/the-great-escape-festival-no-longer-sponsored-by-barclays-following-boycott-43817/

Robinson, O. (2018). Travelling Ayahs of the nineteenth and twentieth centuries: Global networks and mobilization of agency. *History Workshop Journal*, *86*, 44–66. https://doi.org/10.1093/hwj/dby016

Robson, D. (2017, 19 January). How East and West think in profoundly different ways. *BBC Future*. www.bbc.com/future/ article/20170118-how-east-and-west-think-in-profoundly-different-ways

Rohmer, S. (1913). *The Mystery of Dr. Fu-Manchu*. London: Methuen & Co. Ltd.

Roll, M. (2021, October). *Korean Wave (Hallyu) – Rise of Korea's cultural economy & pop culture*. Martin Roll. https://martinroll.com/resources/articles/asia/korean-wave-hallyu-the-rise-of-koreas-cultural-economy-pop-culture/

Roy, D. (2022, 31 October). India's tea estate workers: Designing an empathic behavioural system to motivate migratory behaviour. *South Asia@LSE*. https://blogs.lse.ac.uk/southasia/2022/10/31/indias-tea-estate-workers-designing-an-empathic-behavioural-system-to-motivate-migratory-behaviour/

Royal Geographical Society (with IBG). (2023). *England and Wales are more ethnically diverse and less segregated than ever before.* www.rgs.org/about-us/press-and-media/recent-press-releases/england-and-wales-are-more-ethnically-diverse-and-less-segregated-than-ever-before

Ruiz, N.G., Im, C., & Tian, Z. (2023). *Discrimination experiences shape most Asian Americans' lives*. Pew Research Center. www.pewresearch.org/race-and-ethnicity/2023/11/30/discrimination-experiences-shape-most-asian-americans-lives/

Runnymede Trust. (2021, 31 March). *Sewell Reports: Runnymede responds*. YouTube. www.youtube.com/watch?v=75RcBXlRwlw

Russell, K. (2023, 20 June). (D. Kam, Interviewer) [Personal communication].

Saad, L. F. (2020). *Me and White Supremacy: Combat Racism, Change the World and Become a Good Ancestor*. Blackstone.

Sahar, T. (2023, 7 August). *Interview with Tiara Sahar* (M.-A. Vu Peterson, Interviewer) [Personal communication].

Said, E. (1978). Orientalism. W. Ross Macdonald School, Resource Services Library.

Samshasha. (2015). *Chinese Mythology*. www.glbtqarchive.com/literature/chinese_myth_L.pdf

Sarkar, A. (2025). *Minority Rule: Adventures in the Culture War*. Bloomsbury Publishing.

Sato, M. (2024, 3 April). *Meta's AI image generator can't imagine an Asian man with a white woman*. The Verge. www.theverge.

com/2024/4/3/24120029/instagram-meta-ai-sticker-genera-tor-asian-people-racism

Schouler-Ocak, M., Bhugra, D., Kastrup, M., Dom, G., Heinz, A., Küey, L., & Gorwood, P. (2021). Racism and mental health and the role of mental health professionals. *European Psychiatry, 64*(1), 1–15. https://doi.org/10.1192/j.eurpsy.2021.2216

Scott, S. (2017). The wages of fear: Risk, safety and undocumented work. *Sociology, 51*(5), 1042–1058. www.researchgate.net/publication/303944648_The_Wages_of_Fear_risk_safety_and_undocumented_work

Sen, A. (2006). *Identity and Violence: The Illusion of Destiny.* W.W. Norton & Company.

Shafi, A., & Nagdee, I. (2022). *Race to the Bottom: Reclaiming Antiracism.* Pluto Press.

Shor, E., & Golriz, G. (2018). Gender, race, and aggression in mainstream pornography. *Archives of Sexual Behavior, 48*(3), 739–751. https://doi.org/10.1007/s10508-018-1304-6

Shukla, N. (2023, 19 June). *Personal communication* [Email to Amy].

Sims, J. M. (2006, September). *The Vietnamese Community in Great Britain.* Runnymede Trust. www.runnymedetrust.org/publications/the-vietnamese-community-in-great-britain

Soh, S. (2020, 8 May). In *2001, a British-Chinese protest against virus-related racism.* Al Jazeera. www.aljazeera.com/features/2020/5/8/in-2001-a-british-chinese-protest-against-virus-related-racism

Song, M. (2003). *Choosing Ethnic Identity.* Polity Press.

Soon, W.-C. (2022, July). (D. Kam, Interviewer) [Personal communication].

Southeast and East Asian Centre (SEEAC). (2023). *Not just 'Asian others': Research and outreach toolkits for engagement with East and Southeast Asian communities.* www.seeac.org.uk/blogs/not-just-asian-others-research-and-outreach-toolkits-for-engagement-with-east-and-southeast-asian-communities

Spade, D. (2020). *Solidarity not charity.* The Anarchist Library. https://theanarchistlibrary.org/library/dean-spade-solidarity-not-charity

Stephens, K. (2019, 14 November). *Grieve the Essex 39, but recognise the root causes.* Institute of Race Relations. https://irr.org.uk/article/grieve-the-essex-39-but-recognise-the-root-causes/

Stone, N. (2017). *The Yellow Peril in Britain.* International Institute of Social and Economic Sciences.

Stop AAPI Hate. (2022, 4 March). *National Report (Through December 31, 2021).* Stop AAPI Hate. https://stopaapihate.org/2022/03/04/national-report-through-december-31-2021/

Sussex, K. (2023, 15 June). (M.-A. Vu Peterson, Interviewer) [Personal communication].

Svitak, A. (2014, 5 May). The beauty problem we don't talk about. *HuffPost.* www.huffpost.com/entry/teen-body-image_b_5251604

Syal, R., Sabbagh, D., & Stacey, K. (2023, 30 October). Suella Braverman calls pro-Palestine demos 'hate marches'. *The Guardian.* www.theguardian.com/politics/2023/oct/30/uk-ministers-cobra-meeting-terrorism-threat-israel-hamas-conflict-suella-braverman

Taylor, B., Akoka, K., Berlinghoff, M., & Havkin, S. (2021). *When Boat People were Resettled, 1975–1983: A Comparative History of European and Israeli Responses to the South-East Asian Refugee Crisis.* Palgrave Macmillan, Springer Nature. https://link.springer.com/chapter/10.1007/978-3-030-64224-2_4

Tchen, J.K.W. (2005). *Yellow Peril: An Archive of Anti-Asian Fear.* Harvard University Press.

Thai, B. D. (2007, May). Mixed language: Mixed identities? The Mechanisms of ethnic retention and cultural reproduction in later-generation Vietnamese immigrants in Australia. Proceedings of the Redesigning Pedagogy: Culture, Knowledge and Understanding Conference, Singapore.

The Federation of Small Businesses. (2019, 2 May). *Revealed: Immigrant-led businesses' £13 billion contribution to Scotland.* www.fsb.org.uk/resources-page/revealed-immigrant-led-businesses-13-billion-contribution-to-scotland.html

The Guardian. (2020, 21 June). *Trump uses 'kung flu' to describe coronavirus at rally, drawing criticism from civil rights groups.* www.theguardian.com/us-news/2020/jun/20/trump-covid-19-kung-flu-racist-language

The Guardian. (2024a, 30 May). *Edinburgh international book festival ends Baillie Gifford sponsorship amid protest threats*. www.theguardian.com/books/article/2024/may/30/edinburgh-international-book-festival-ends-baillie-gifford-partnership

The Guardian. (2024b, 7 December). *Agency that brought heavily indebted Indonesian workers to UK loses licence*. www.theguardian.com/uk-news/2024/dec/07/agency-that-brought-heavily-indebted-indonesian-workers-to-uk-loses-licence

The Migration Observatory. (2024). *Migrants in the UK labour market: An overview*. https://migrationobservatory.ox.ac.uk/resources/briefings/migrants-in-the-uk-labour-market-an-overview/

The Monitoring Group. (n.d.). *About 1*. https://tmg-uk.org/historyoftmg

The Old Bailey Proceedings Online Project. (2003). *The proceedings of the Old Bailey*. www.oldbaileyonline.org/about/chinese

The University of Edinburgh. (2023, 21 November). *University factsheet*. https://governance-strategic-planning.ed.ac.uk/facts-and-figures/university-factsheet

The Voice of Domestic Workers. (n.d.). *Our campaign*. www.thevoiceofdomesticworkers.com/our-campaign

Tian, N. G. R., Im, C., & Ziyao, T. (2023, 30 November). *Asian Americans and the 'Forever Foreigner' stereotype*. Pew Research Center. www.pewresearch.org/2023/11/30/asian-americans-and-the-forever-foreigner-stereotype/

Tingvold, L., Hauff, E., Allen, J., & Middelthon, A. L. (2012). Vietnamese refugee parenting practices and adolescent well-being: A qualitative study. *International Journal of Intercultural Relations, 36*(4), 563–574.

Townsend, M. (2023, 5 March). UK government 'complicit' in asylum seeker hotel attacks, say unions. *The Observer*. www.theguardian.com/uk-news/2023/mar/05/uk-government-complicit-asylum-seeker-hotel-attacks-say-unions

Trần, B. T. (2014). *The Red Earth*. Ohio University Press.

Turkle, S. (2015). *Reclaiming Conversation: The Power of Talk in a Digital Age*. Penguin Press.

UCL. (2014). *Positive economic impact of UK immigration from the European Union: New evidence*. UCL News. www.ucl.ac.uk/

news/2014/nov/positive-economic-impact-uk-immigration-european-union-new-evidence

UK Data Service. (2017, June). *The impact of migrants on London, its workforce and its economy.* https://ukdataservice.ac.uk/case-study/the-impact-of-migrants-on-london-its-workforce-and-its-economy/

University of Leicester. (2024, 26 April). *Rising hate crimes against ESEA communities in the UK: A University of Leicester report.* https://le.ac.uk/news/2024/april/race-study

University of Nottingham Rights Lab. (2021). *Comparative Report: the top 20 source countries for Modern Slavery Victims in the UK.* www.nottingham.ac.uk/research/beacons-of-excellence/rights-lab/resources/reports-and-briefings/2021/april/the-top-20-source-countries-for-modern-slavery-in-the-uk.pdf

Vincent, C. (2019). Cohesion, citizenship and coherence: Schools' responses to the British values policy. *British Journal of Sociology of Education, 40*(1), 17–32. https://doi.org/10.1080/01425692.2018.1496011

Vu, F. (2023, September). *Interview with Frankie Vu* (I. Pan, Interviewer) [Personal communication].

Vu Peterson, M.-A. (2021, 11 July). Chats on mixed identity [Podcast episode]. In *But Where Are You From?* Spotify. https://open.spotify.com/episode/3ynOAUWSFYUmdm9PFoJyhO?si=3968d2489e5c479a

Vu Peterson, M.-A. (2022, 13 March). Why aren't you BROWN? [Podcast episode]. In *But Where Are You From?* Spotify. https://open.spotify.com/episode/3EBU31auBqjYs8C0YrGojp?si=WuVqncIVQ0yx1m9a2ieTcA

Ward, D., Lawrence, F., Carter, H., Watts, J., Dodd, V., & Pai, H.-H. (2004, 7 February). Victims of the sands and the snakeheads. *The Guardian.* www.theguardian.com/uk/2004/feb/07/china.immigration1

Warner, A. (2021a, 29 March). *Storytelling through travel, with Alicia Warner* (M.-A. Vu Peterson, Interviewer) [Interview]. besea.n. www.besean.co.uk/spotlight/alicia-warner

Warner, A. (Director). (2021b, September). *Land, Sea & Stars* [Film].

Watson, J. (2004). Presidential address: Virtual kinship, real estate, and diaspora formation –The man lineage revisited. *The Journal* of Asian Studies, 63(4), 893–910. https://doi.org/10.1017/s0021911804002359

Wellman, M. L. (2022). Black squares for Black lives? Performative allyship as credibility maintenance for social media influencers on Instagram. Social Media + Society, 8(1). https://doi.org/10.1177/20563051221080473

Wikipedia Contributors. (2019, 26 May). Prostitution statistics by country. Wikipedia. https://en.wikipedia.org/wiki/Prostitution_statistics_by_country

Wilkins, A. (2021). Migration, work and home-making in the city: Dwelling and belonging in Vietnamese communities in London. Journal of the Royal Anthropological Institute, 27(4), 1020–1021. https://doi.org/10.1111/1467-9655.13635

Wintour, P. (2004, 26 February). Whip removed from Tory MP over race joke. The Guardian. www.theguardian.com/politics/2004/feb/26/conservatives.immigrationpolicy

Witchard, A. (2017). Our migration story: Chinese Limehouse and Mr Ma and Son. www.ourmigrationstory.org.uk/oms/chinese-limehouse-and-mr-ma-and-son

Woolfson, D. (2021). Ice cream 2021: Little Moons is 900% hero of handheld. The Grocer. www.thegrocer.co.uk/analysis-and-features/ice-cream-2021-little-moons-is-900-hero-of-handheld/662840.article.

Wu, K. (2024, 25 February). WhatsApp discussion on identity (WhatsApp to Amy) [Personal communication].

Xu, G. (2011). Strangers on the Western Front: Chinese Workers in the Great War. Harvard University Press.

Yeh, D. (2018). The cultural politics of in/visibility: Contesting 'British Chineseness' in the arts. In D. Yeh & A. Thorpe (Eds.), Contesting British Chinese Culture. Palgrave Macmillan. pp. 31–59.

Yeh, D. (2021). Becoming British East Asian and Southeast Asian. British Journal of Chinese Studies, 11. https://bjocs.site/index.php/bjocs/article/view/131

Yeh, D. (2023, 9 August). Interview with Diana Yeh (K.-W. Hii, Interviewer) [Personal communication].

Yehuda, R., & Lehrner, A. (2018). Intergenerational transmission of trauma effects: Putative role of epigenetic mechanisms. World Psychiatry, 17(3), 243–257. https://doi.org/10.1002/wps.20568

Yoo, H.C., Burrola, K.S., & Steger, M.F. (2010). A preliminary report on a new measure: Internalization of the Model Minority Myth Measure (IM-4) and its psychological correlates among Asian American college students. *Journal of Counseling Psychology, 57*(1), 114–127.

YouGov. (2024, 9 September). *Should Britain allow people fleeing persecution or war in other countries to come and live in Britain?* https://yougov.co.uk/topics/politics/trackers/should-britain-allow-people-fleeing-persecution-or-war-in-other-countries-to-come-and-live-in-britain

Zhang, S. (2021, 27 December). *Between Soho and Chinatown.* Queer China UK. https://queerchinauk.com/2021/12/27/between-soho-and-chinatown-sj-zhang/

Zhou, L. (2021, 5 May). *The inadequacy of the term 'Asian American.'* Vox. www.vox.com/identities/22380197/asian-american-pacific-islander-aapi-heritage-anti-asian-hate-attacks

Zhu, Y. (2023). The ladder of heritage interpretation. In N. A. Silberman (Ed.), *An Anthology of World Heritage Interpretation and Presentation.* Chae Su-hee (Director General, WHIPIC).

INDEX